HELMETS OF THE FIRST WORLD WAR

HELMETS OF THE FIRST WORLD WAR
Germany, Britain & their allies

Michael J Haselgrove & Branislav Radovic

Schiffer Military History
Atglen, PA

THE PHOTOGRAPHS

The overwhelming majority of period photographs used in this book are sourced and reproduced by kind permission of the Imperial War Museum, who retain the copyright. Viewing the huge IWM archive is by appointment only, and copies for personal use can be obtained for a nominal fee (copies for publication are subject to a separate tariff). Enquiries should be addressed to:

Photograph Archive
Imperial War Museum
Lambeth Road, London SE1 6HZ
United Kingdom
Telephone: 0207 416 5333
E-Mail: photos@iwm.org.uk
Fax: 0207 416 5355

The IWM numbers quoted after each image refer to each photograph's negative number in the archive – eg. Q 123456.

Book Design by Branislav Radovic and Robert Biondi.
Photography by Robert Biondi.

Library of Congress Catalog Number: 99-67579.

Printed in China.
ISBN: 0-7643-1020-8

Published by Schiffer Publishing Ltd.
4880 Lower Valley Road
Atglen, PA 19310 USA
Phone: (610) 593-1777
FAX: (610) 593-2002
E-mail: Schifferbk@aol.com.
Visit our web site at: www.schifferbooks.com
Please write for a free catalog.
This book may be purchased from the publisher.
Please include $3.95 postage.
Try your bookstore first.

In Europe, Schiffer books are distributed by:
Bushwood Books
6 Marksbury Ave.
Kew Gardens
Surrey TW9 4JF
England
Phone: 44 (0)208 392-8585
FAX: 44 (0)208 392-9876
E-mail: Bushwd@aol.com.
Free postage in the UK. Europe: air mail at cost.
Try your bookstore first.

Contents

Introduction

This book is intended as an appreciation of helmets of the First World War. However, in reality it is far more than that. The reader will find numerous photographs in full colour showing, in great detail, both exterior and interior views of many helmets. The authors believe that some of the detail shown has not been reproduced in any previous publication. Subjects covered in this book include helmet shell types, paint finishes, insignia both painted and applied, manufacturers' marks, liner systems, camouflage finishes and covers.

We are conscious that it is becoming increasingly difficult to acquire genuine equipment, including helmets, from the First World War period. There are, unfortunately, many items on the market that are either fake or that have been restored or altered in some way. Many of the best pieces are in museums or private collections where it is hard for the average would-be collector to gain hands-on experience. Whereas there is no real substitute for actually handling an original piece, we hope that the close-up photographs shown here will assist in identification and help direct the collector or student towards the original and away from the dubious. However, we sound a note of caution. No helmet should be rejected just because you cannot find it, or a feature that it has, in this or any other book. Rare experimental helmets or those that bear contemporary modification may still be encountered. Personal experience and judgment have to be exercised at all times.

Having said that, this is not intended to be an in-depth reference book. Not all helmets and their variations are shown. The reason for this is that those shown are without exception, in the collections of the authors. No attempt has been made to include photographs of items from museums or other collections. For those who require further reading or more information, a list of suggested books and publications is included at the end of the book.

The helmets are presented in sections. We have included at the end of each section a number of photographs taken during the war. For the most part, they have been gleaned after many hours of search from the Department of Photographs at the Imperial War Museum in London. We are grateful for permission to reproduce them here. In some cases a complete photograph is shown, whilst in others we only show a "crop" to illustrate a particular helmet. The photographs speak for themselves.

The only comment we make is that it was possible to find quite a number showing German troops wearing camouflage painted helmets. Prior to our search we had thought that such photographs were extremely rare.

We hope that the photographs and information contained in this book add, in some small way, to the knowledge of the reader and that we have managed to impart something of our enthusiasm and fascination for the subject.

GERMANY

Model 1895 Prussian pickelhaube. Black leather with gilt brass fittings. Manufactured by Alexander Dahl Helm-Fabrik Barmen in 1913.

T.R.8.
1913

Model 1915 Prussian pickelhaube. Black leather with matte grey finished steel fittings. Manufactured by Alexander Dahl Helm-Fabrik Barmen in 1916.

58
1916

Prussian ersatz pickelhaube. It has a one-piece felt body with grey finished steel fittings. Produced circa 1916.

Prussian ersatz pickelhaube. Introduced in late 1914. It has a three-piece metal body painted with black gloss enamel. The fittings are gilt-brass. This helmet was manufactured in 1915.

Prussian ersatz pickelhaube. Introduced in late 1914. It has a one-piece metal body painted field grey with brass fittings.

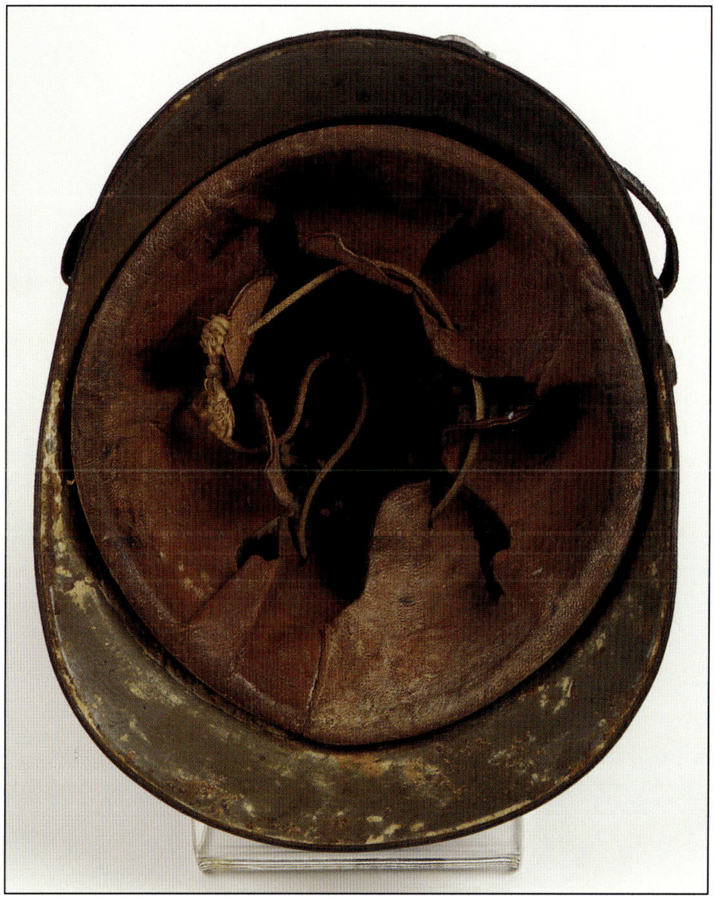

■ Pickelhaube cover. The number denotes the regiment, in this case the 55th Infantry Regiment Graf Bülow von Dennewitz, 6th Westfalishes. This is a good quality pre-war enlisted man's cover. Shortly after the war started regimental markings were changed to green. By the end of 1915 the use of regimental markings was largely discontinued.

■ Pre-war pickelhaube cover. The number, in this case for the 119th Grenadier Regiment Konigin Olga, 1st Würtembergisches. The R was used to denote Reserve units.

55

Wartime enlisted man's pickelhaube cover. Note the coarse material and absence of regimental markings.

Prussian shako worn by Jäger troops and the communications services. Leather construction with grey finished steel fittings. This helmet was manufactured by Alexander Dahl Barmen. Circa 1916.

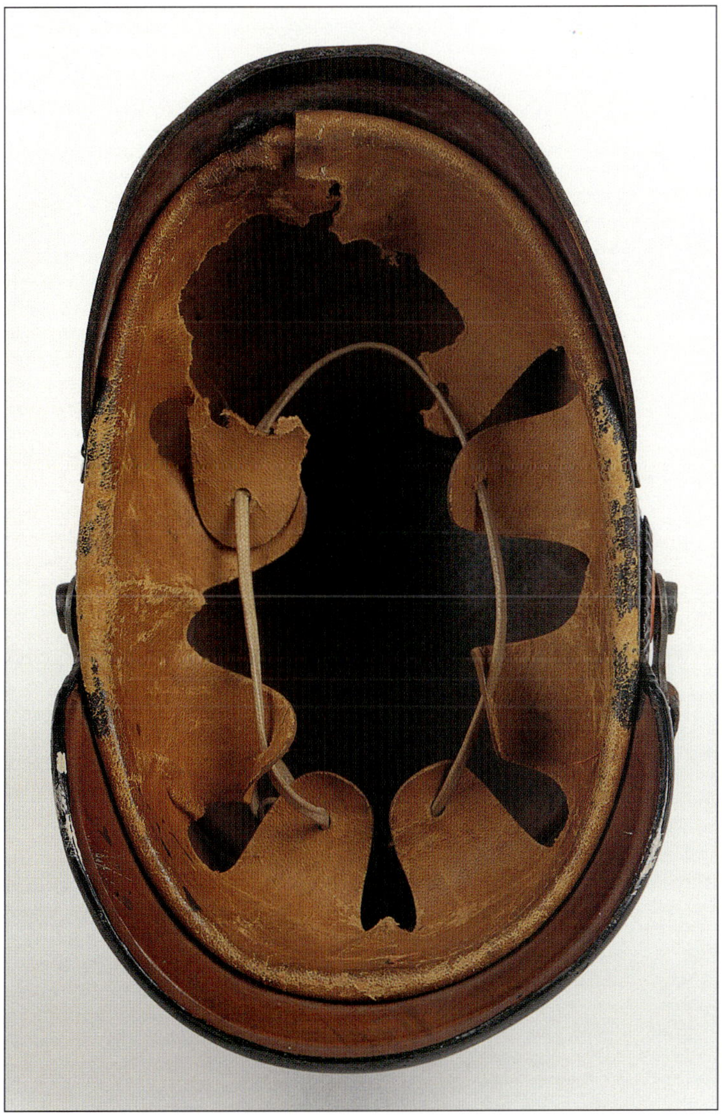

Wartime enlisted man's Prussian Uhlan Tschapka. It has a leather body with black enamel painted steel top. Grey finished steel fittings. Shown on the right with its cover. The cover is fastened by a series of hooks and eyes down one side and secured to the helmet by two frontal clips and an elasticated rear.

Model 1916 field grey steel helmet (stahlschutzhelm). Designed by Kapitan Friedrich Schwerd of the Hannover Technical Institute and first issued to the 1st Assault Battalion in December 1915 for evaluation. The first issue to frontline troops was at Verdun at the end of January 1916. This type is known as the "square dip visor" and is thought to be the first version of the Model 1916.

Model 1916 field grey steel helmet (production model).

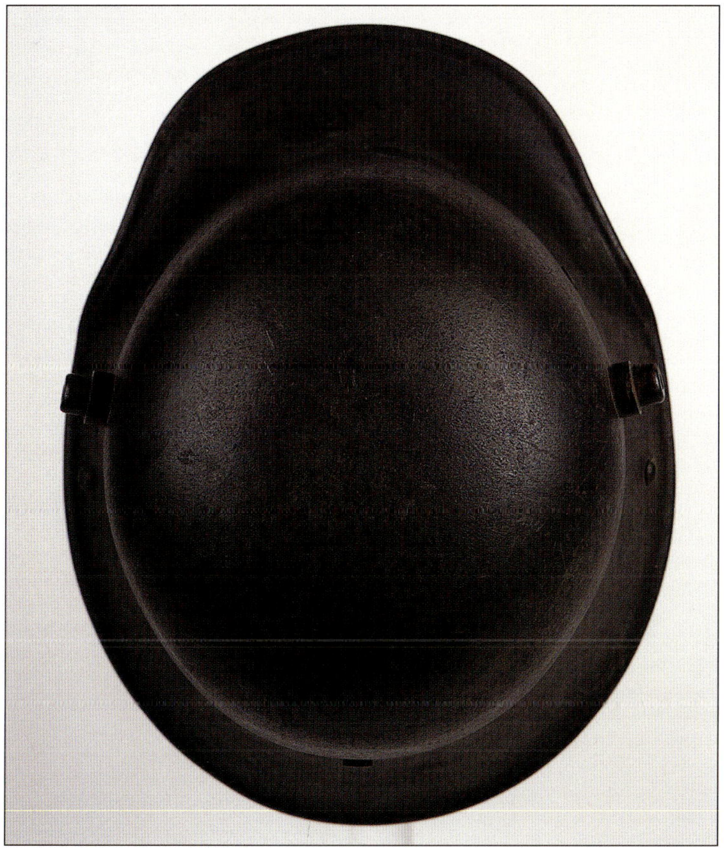

Model 1916 interior. Tan leather liner pads and liner band. Model 1891 lugs to attach the one-piece chinstrap.

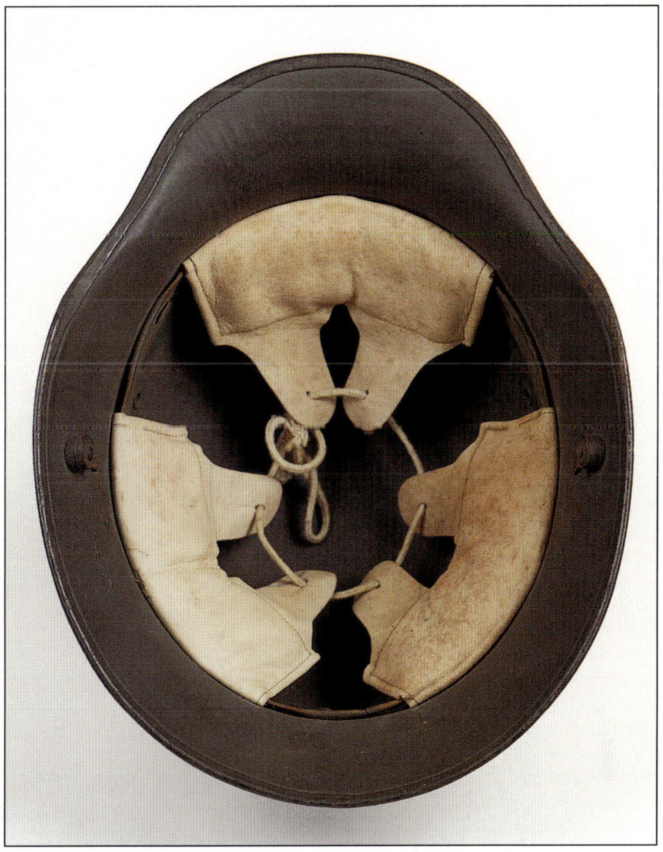

Model 1917 interior. Tan leather liner pads with metal liner band. Externally the Model 1917 helmet is identical to the Model 1916.

Model 1917 interior. White leather liner pads.

Model 1918 interior. White leather liner pads with metal band incorporating the new two-piece chinstrap attached directly onto the liner band.

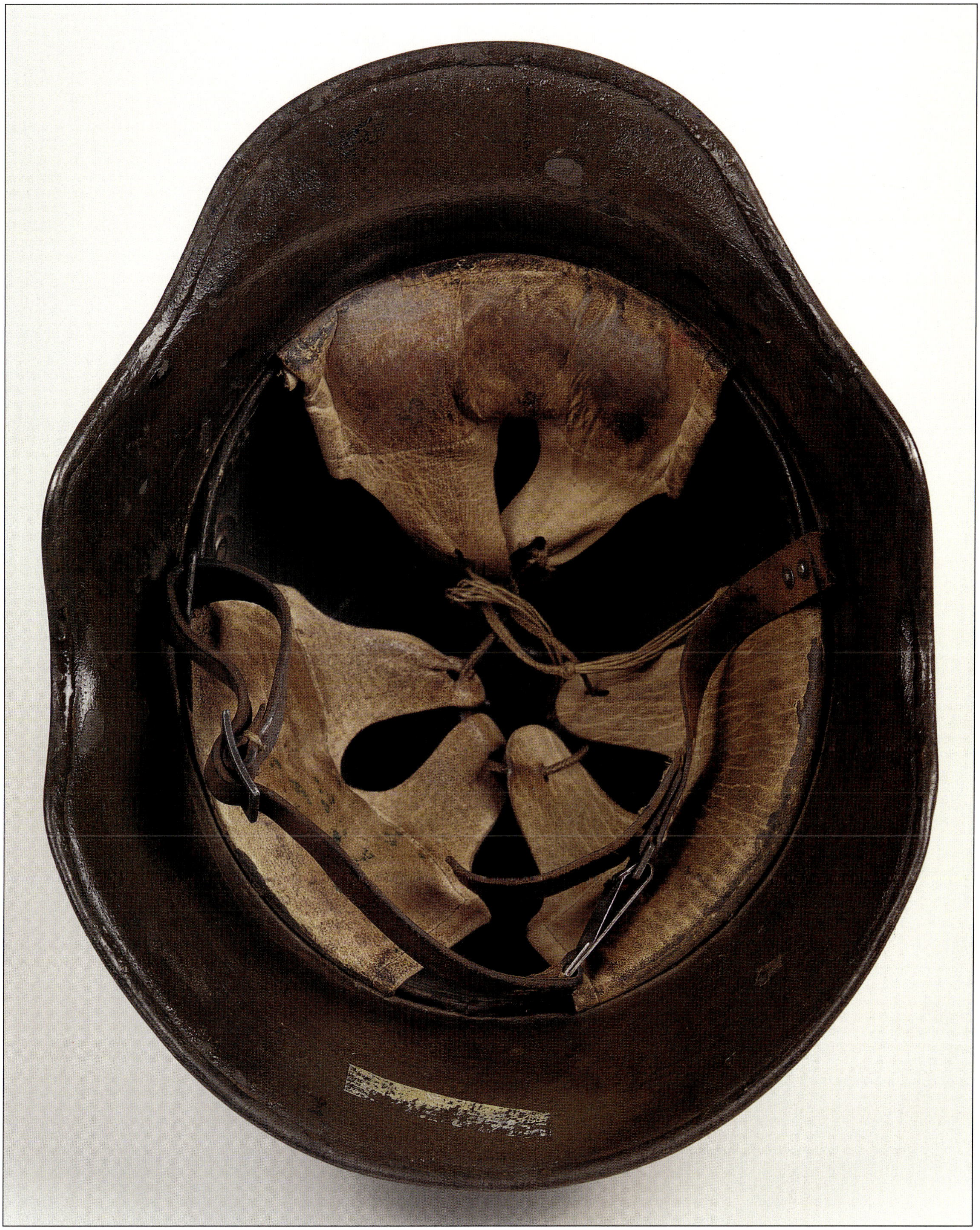

Model 1918 cut-out interior. White leather liner pads with metal band incorporating the new two-piece chinstrap attached directly onto the liner band. The cut-out modification was an effort to improve hearing for the wearer.

Model 1918 field grey steel helmet.

Model 1918 cut-out steel helmet in field grey finish. These helmets were also produced with a matte felt-like anti-reflective finish.

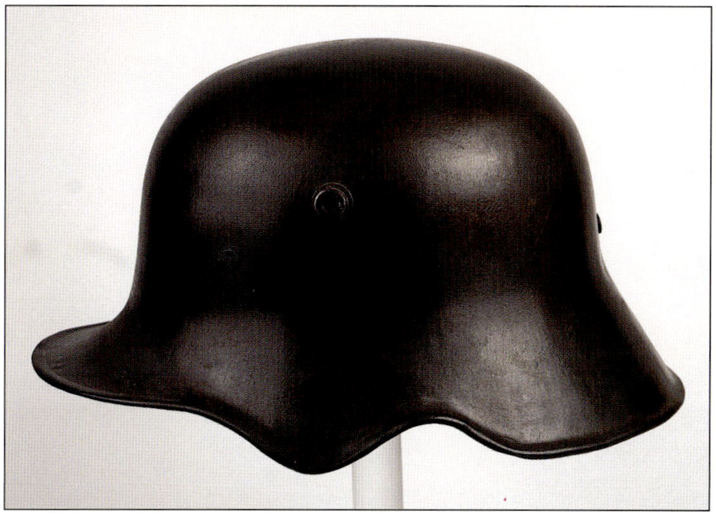

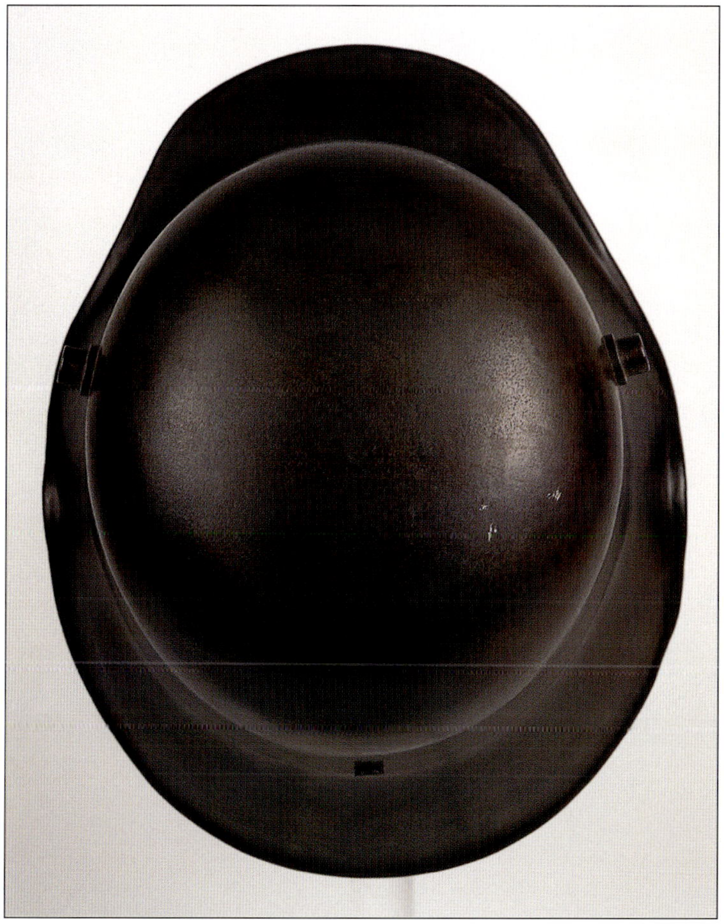

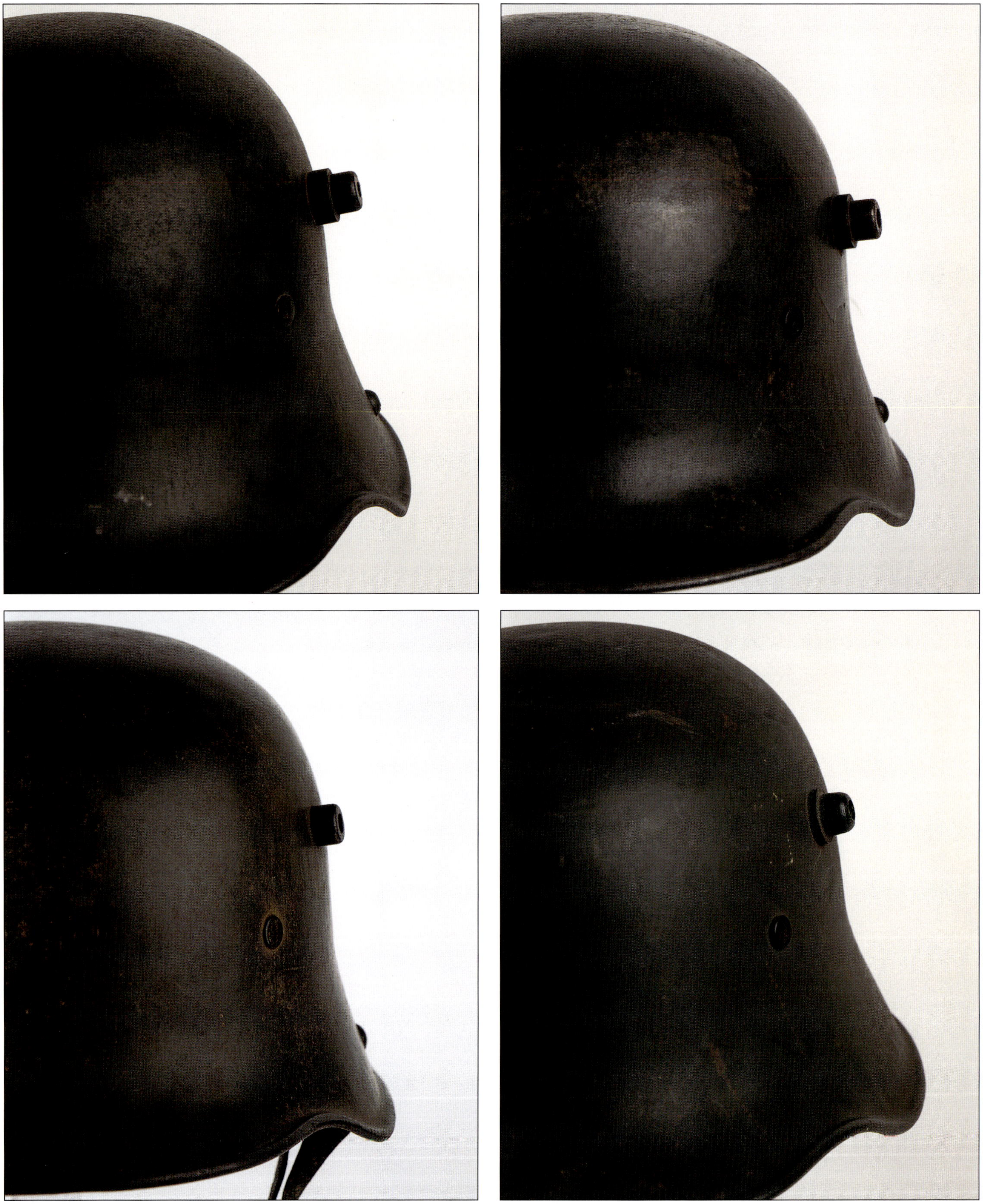

Models 1916-18. Details of the ventilation lugs that support the armoured brow plate (stirnpanzer). The smaller the helmet size, the deeper the step in the lug required to ensure a good fit for the brow plate. Shown here (clockwise from the top left) are the variations in size: 60, 62, 64, 66. Size 66 and 68 (not shown) have identical ventilation lugs.

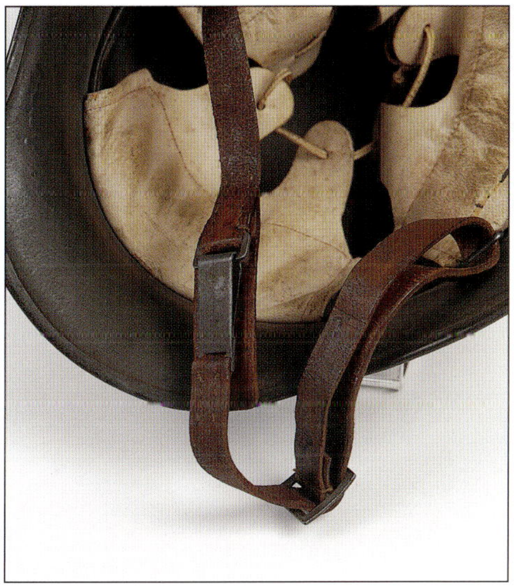

■ Models 1916/17 chinstrap. Model 1891 lug detail (left) and sliding buckle (right) both shown here on a Model 1916.

■ Model 1918 chinstrap. Liner attachment detail (left), two-piece carbine clip (centre), and clip cover showing manufacturer's mark (right).

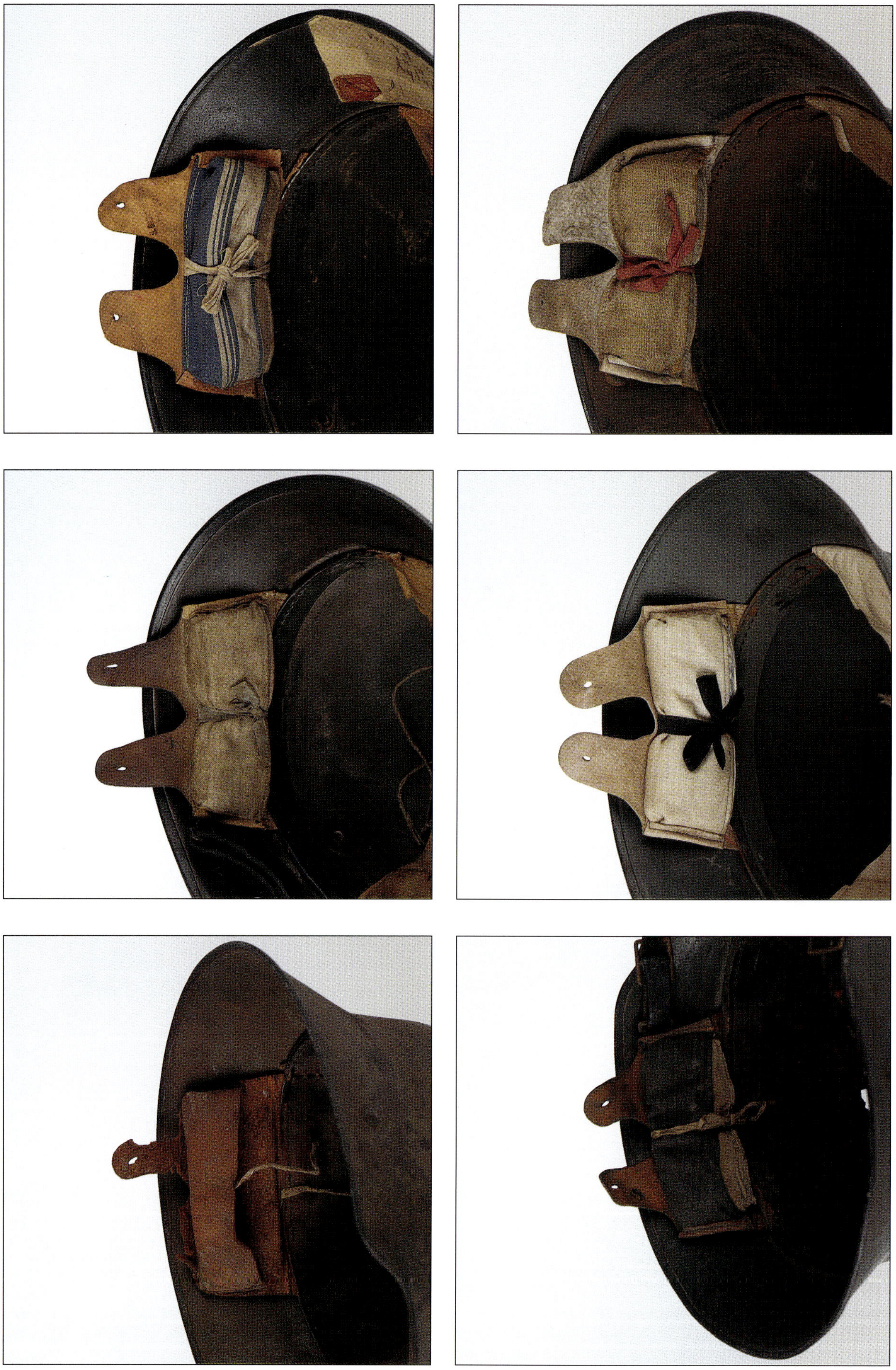

Examples of cloth liner pad pouches found in models 1916, 1917 and 1918.

Examples of horsehair-filled cloth cushion pads for liner pouches.

Various examples of manufacturers' marks and shell size stampings. It is alleged that a size 70 was produced, but to date no example has been encountered by the authors.

B.F 64

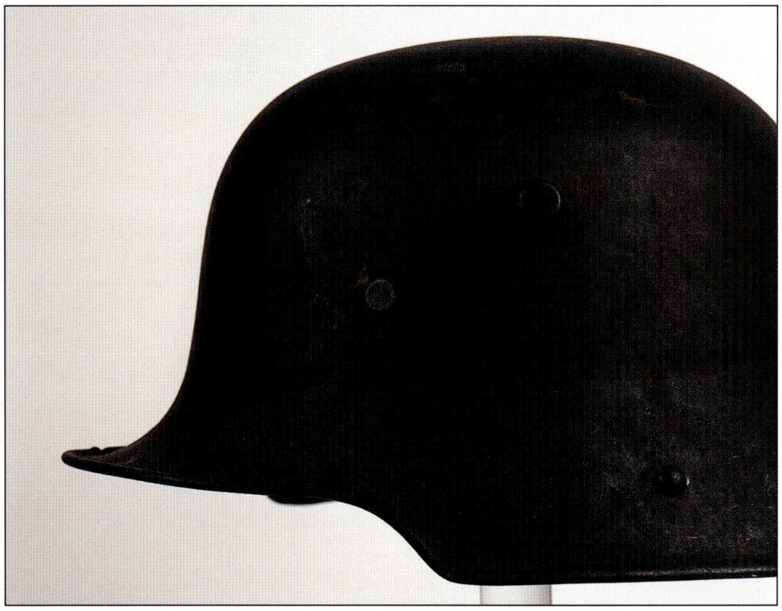

Comparisons of visor depth produced by various manufacturers. The helmet shown at the top is a Model 1916 with a square dip visor.

Model 1916 helmet with variant chinstrap lugs placed on the exterior of the shell.

Model 1916 helmet showing regulation camouflage pattern as laid out in Order II. No. 91 366 and signed by Ludendorff on 7th July 1918. It stipulated that helmets should be painted in several colours, all in keeping with the relevant season. In summer for example: green, yellow ochre and rust brown, separated by a finger-wide black line. Many helmets are found that do not comply with the order, presumably due to individual painting skill and preference.

Model 1916 camouflage variation. This helmet belonged to Leut. Beiman (see section on named helmets, pages 112-113).

Model 1916 camouflage variation. This helmet belonged to Ltn. d.R. Schumann (see section on named helmets, pages 112-113).

Model 1916 camouflage variation.

Model 1916 medic's helmet with hand-painted Gardekorps insignia.

Model 1916 camouflage helmet with hand-painted Gardekorps insignia.

Model 1916 camouflage variation. Possibly a factory applied camouflage scheme.

Model 1916 camouflage variation.

Model 1916 camouflage variation without regulation black dividing lines.

Model 1916 camouflage variation without regulation black dividing lines.

Model 1916 camouflage variation.

Model 1916 camouflage variation showing stress marks from poor shell pressing.

Model 1916 camouflage with hand-painted Prussian eagle insignia to front. Note blocked ventilation holes, possibly to prevent heat loss or wind noise.

Model 1916/17 camouflage helmet shell with hand-painted skull and crossbones insignia to front. The skull and crossbones motif was sometimes used by tank troops or flamethrower units.

Model 1916 camouflage variation.

Model 1916 camouflage variation.

Model 1917 camouflage variation.

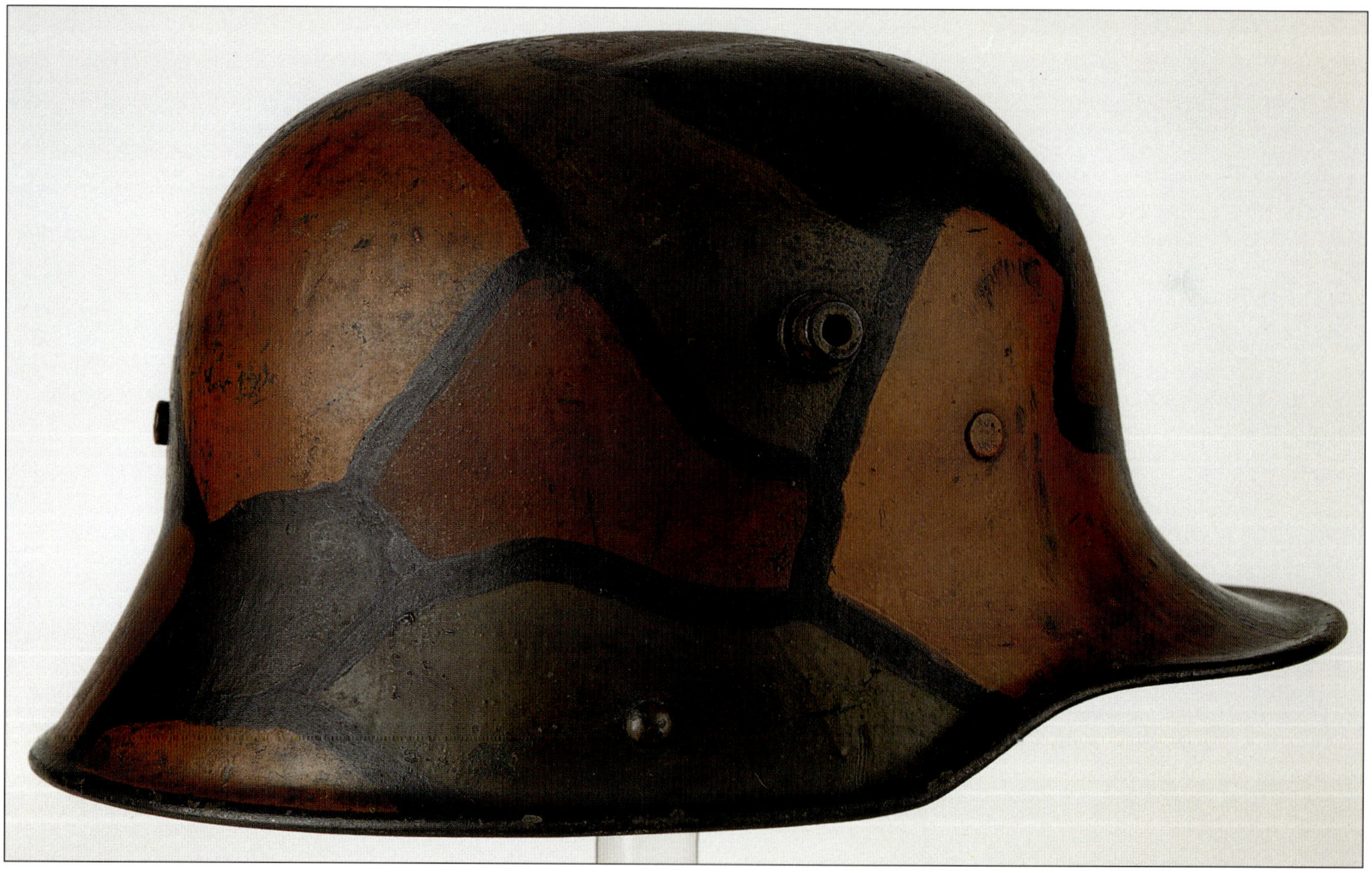

Model 1917 camouflage variation.

Model 1917 fitted with a double liner band and with a camouflage scheme applied over an earlier one.

Model 1917 camouflage variation.

■ Model 1917 camouflage variation.

Model 1917 camouflage variation without regulation black dividing lines.

■ Model 1917 camouflage variation.

Model 1917 camouflage variation.

Model 1917 camouflage variation with British post-home stamps (see pages 124-125).

Model 1917 camouflage variation.

Model 1917 camouflage variation.

Model 1917 camouflage variation.

Model 1917 camouflage variation.

Model 1918 camouflage variation.

Model 1918 camouflage variation.

Model 1918 camouflage variation with bird shape incorporated into the regulation colour scheme.

Model 1918 cut-out camouflaged helmet shell with unit insignia (possibly a company number) applied in white paint.

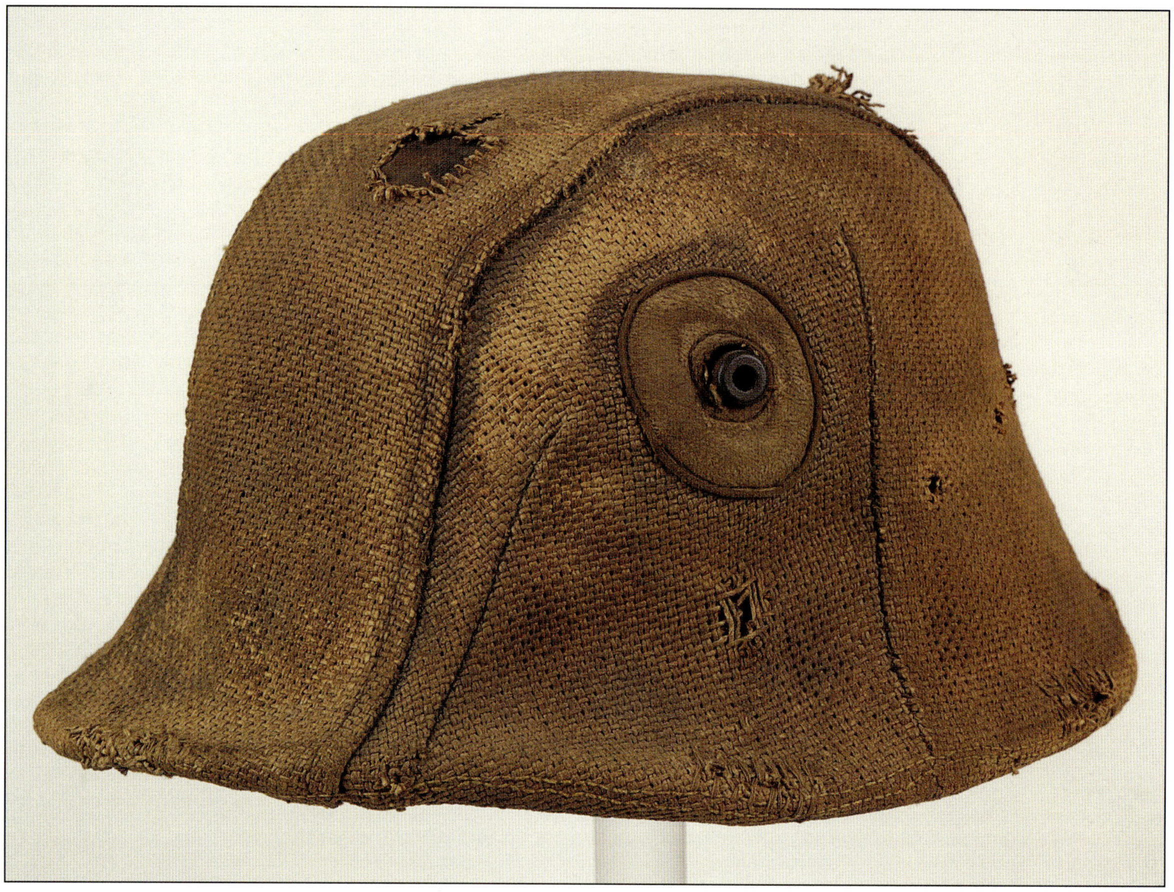

■ Coarse hessian sandbag material drawstring helmet cover in three pieces with leather ventilation lug reinforcing.

■ Tan coloured two-piece helmet cover with leather ventilation lug reinforcing and twisted paper drawstring.

■ Simple two-piece tan material helmet cover with telegraph wire drawstring.

■ Regulation pattern three-piece helmet cover with leather lug reinforcing. This example is in white and has the manufacturer's mark and date printed inside. Variations are found with both 7 and 8 clips sewn onto the cover for attachment to the helmet rim.

Battle-damaged Model 1916/17 camouflage helmet shell.

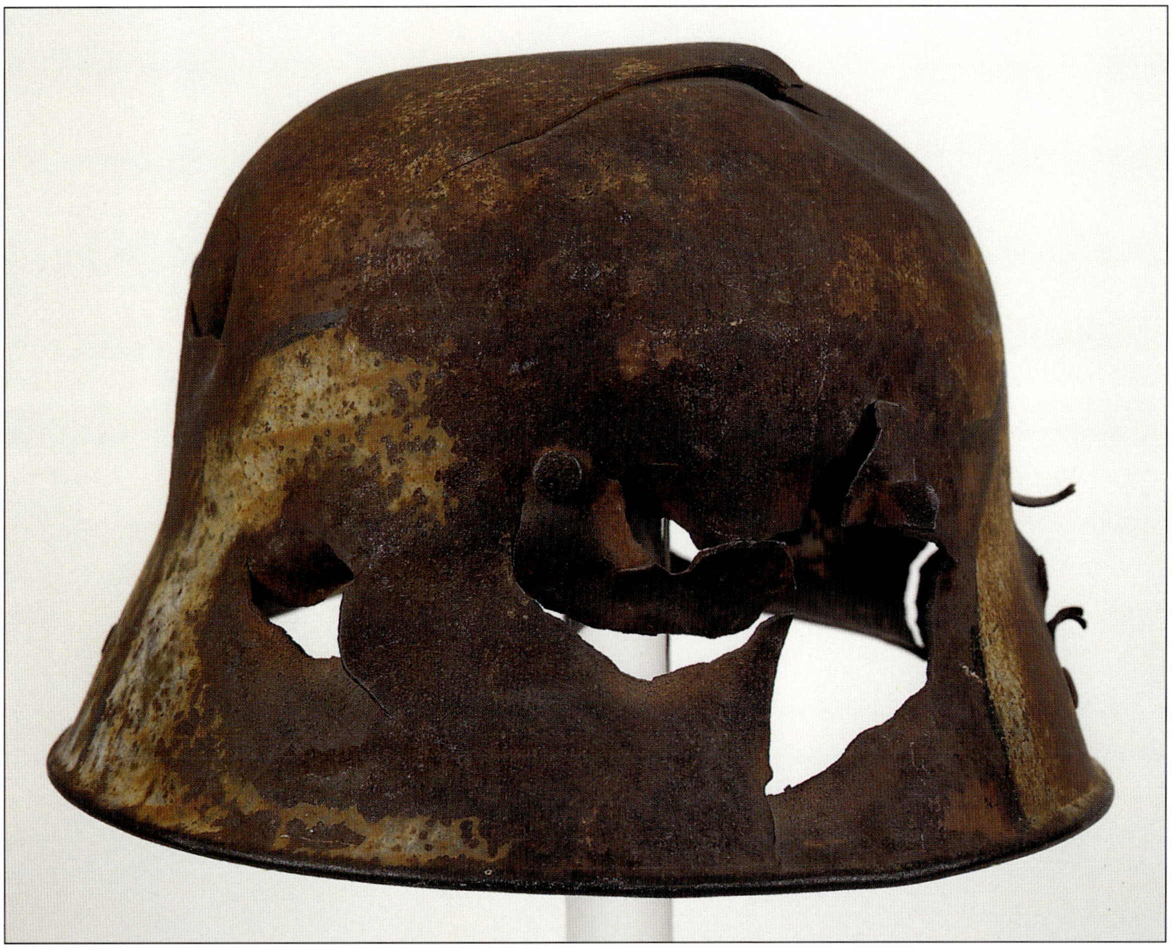

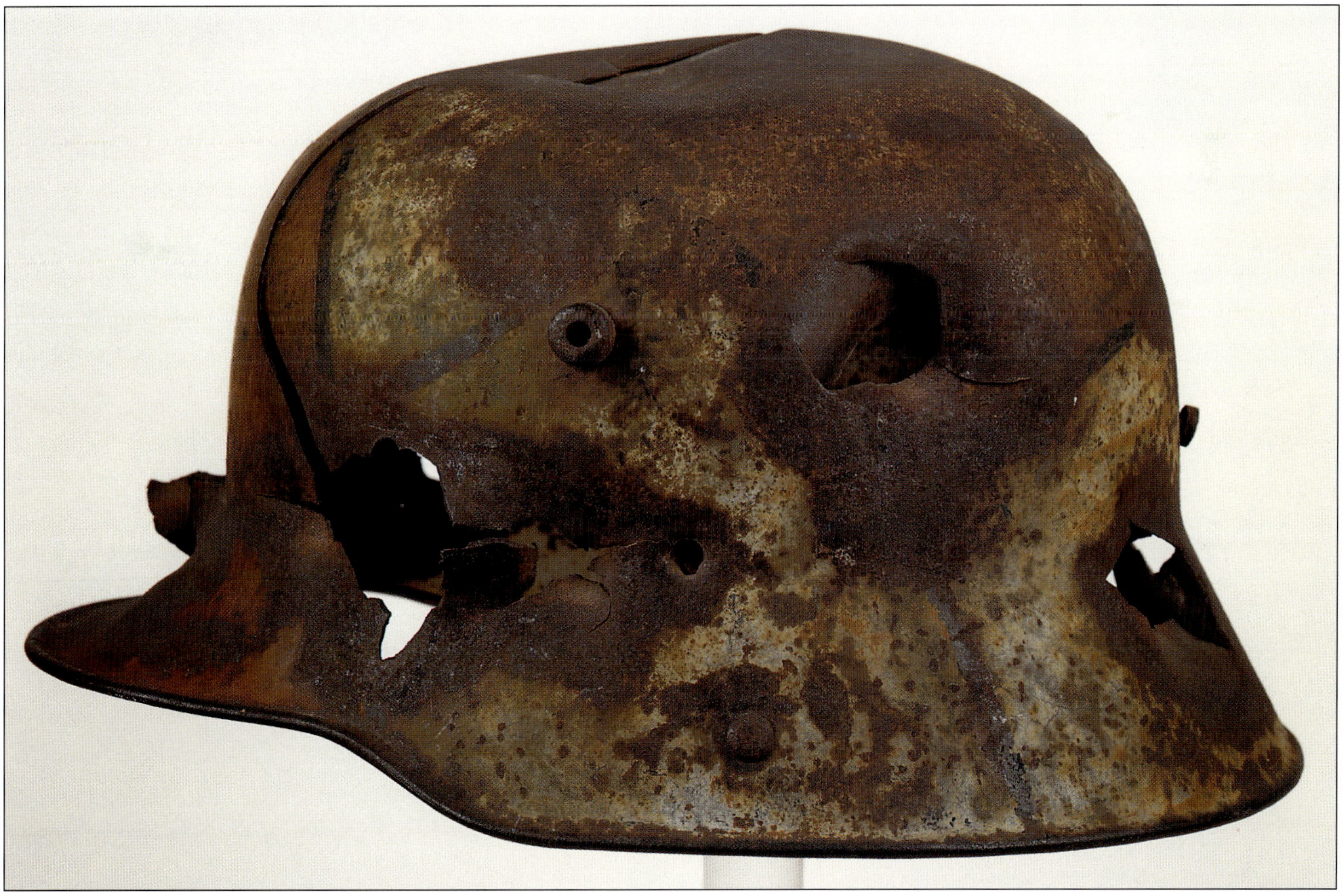

Battle-damaged model 1916 helmet.

Various battle-damaged helmets.

Examples of named helmets.

Model 1916 field grey helmet with armoured brow plate (stirnpanzer) fitted. The use of the brow plate was foreseen almost as early as the helmet itself and was issued to frontline troops at Verdun at the same time in January 1916.

■ Model 1916 field grey helmet with an armoured brow plate showing evidence of bullet strike.

A relic armoured brow plate showing the remains of hessian sandbag and wire camouflage.

An armoured brow plate pierced by small-arms fire.

Several armoured brow plate manufacturers' marks with strap and rivet details.

Early pattern body armour with helmet and brow plate.

Details showing reverse strap, padding and stitching.

■ Late pattern battle-damaged body armour with helmet and brow plate.

■ Details showing rifle butt support and carrying hooks.

Early pattern upper body armour part showing battle damage.

■ Model 1917 helmet with U.S. post-home address label and stamps.

■ Model 1916 helmet with U.S. post-home stamps showing battle damage.

Various examples of post-home trophies. Example at top right was posted back to Britain and is also shown in the camouflage section. The example at bottom left (also shown in the camouflage section) was picked up by Lt. Charles E. Schnell of the U.S. 37th Engineers. Scratched onto the inside brim are the words, "Boche helmet Chatel Chehery October 1918".

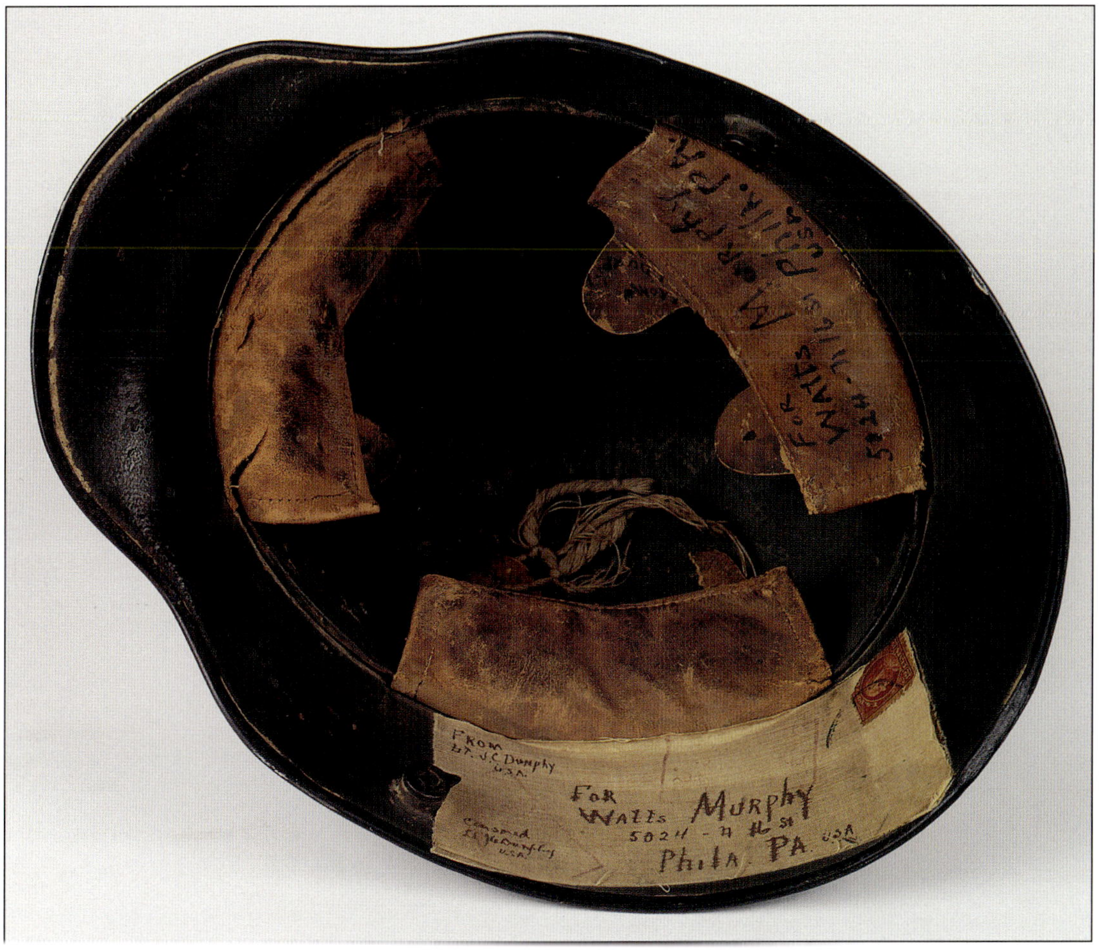
For
Watts Murphy
5024 - N 11th st
Phila. PA. U.S.A.
FROM
Lt. J.C. Dunphy
U.S.A.
For
Watts Murphy
Phila. PA

- Model 1916 helmet trophy with silver plaque inscribed, "Les Boeufs September 1916".

- Model 1917 helmet trophy with painted inscription, "German Trench Helmet Souvenir of the Great War. Le Quesnoy-Belgium. Nov. 4th 1918".

German troops wearing a mixture of headgear: pickelhaubes with covers and Model 1916 steel helmets. Circa 1916. (IWM CO 23761)

(Left) Model 1895 Pre-war Prussian pickelhaube with NCO's chinstrap. (Centre) A Prussian wartime ersatz one-piece metal pickelhaube, worn here by Günther Wagner, Cottbus, April 1915. (Right) A Canadian soldier collects his trophies; an armful of model 1895 pickelhaubes, Cambrai, October 1918. (IWM Q 3383)

British troops wearing a mixture of captured Prussian, Baden, Württemberg and Guards pickelhaubes after the Somme offensive, July 1916. (IWM Q 83)

■ (Left) A pickelhaube clip on canvas cover with central seam. (Right) A plain canvas clip on cover.

■ (Left) Examples of clip on cloth covers with faint stenciled regimental numbering. (Right) A soldier from the 1st Battalion Infantry Reserve Regiment 250, 2nd Company, 75th Reserve Division, wearing a numbered cloth cover.

A crew manning a 17cm minenwerfer, the soldier on the left wears a square dip visor Model 1916, very different in shape from his comrades' helmets. (IWM Q 23784)

(Left) An MG08 machine gun crew member wearing a square dip visor Model 1916 helmet, Western Front. (IWM Q 5034) (Right) Another example of the square dip visor being worn. (IWM Q 496)

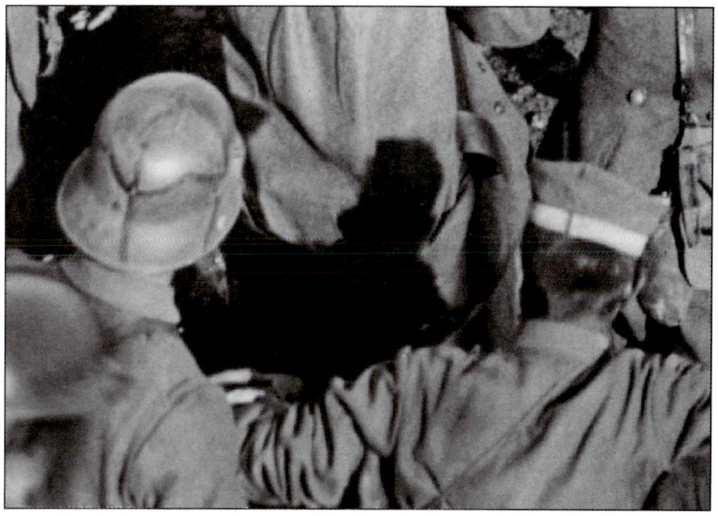

■ (Left) Newly-taken prisoners of war wearing camouflaged Model 1916/17 helmets. East of Arras. October 1918. (Right) Prisoners of war captured during the Battle of the Canal du Nord with camouflaged Model 1916/17 helmets. (IWM Q 9328)

■ (Left) Captured Germans wearing camouflaged Model 1916/17 helmets, Moeuvres. September 1918. (Right) An artillery unit wearing camouflaged helmets with an unknown shield insignia, 1918. (IWM Q 9345)

■ (Left and right) German soldiers wearing brightly camouflaged Model 1916/17 helmets. They were captured during the British attack on Moyennville. August 1918. (IWM Q 6981)

■ A German trench raiding party with their prisoners. Some soldiers wear plain canvas drawstring helmet covers. Flanders. June 1916.

■ (Left) German prisoners of war wearing plain sacking drawstring helmet covers. Note the variation of fitted and unfitted types. Hooge. September 1917. (IWM E(Aus) 799) (Right) A German stretcher-bearer wearing a sacking drawstring cover. Battle of the Aisne near Noyon and Montdidier. June 1918. (IWM Q 55013)

German column of reserves wearing various cloth helmet covers. The NCOs at the head of the column wear the regulation clip-type cover. They are marching along the Albert road towards the front line. March 1918. (IWM Q 51466)

(Left) A German stretcher-bearer wears a sacking cover. Possibly the same man shown on the previous page. (IWM Q 55014) (Right) More examples of German prisoners wearing various types of sacking helmet covers. Hooge. September 1917. (IWM E(Aus) 799)

■ The armoured brow plate being worn without the body armour. Western Front. (IWM Q 107458)

■ (Left) Armoured brow plates seen here being worn by men of the 8th Company, 2nd Battalion Landwehr Infantry Regiment 85. Argonne. October 1916. (Courtesy of Greg Nevill) (Right) The brow plate worn with body armour. (IWM Q 50348)

A posed photograph of a German stormtrooper wearing an early pattern body armour and grenade bags. The photograph is captioned as 1917.

■ German troops wearing Model 1916/17 helmets, brow plate and early pattern body armour. Western Front. (IWM Q 50348)

■ The figure on the right wears body armour. These men were captured by Canadians near Lens. September 1917. (IWM CO 1966)

■ Irish Guardsmen inspect a captured MG08 with its armoured jacket. All wear captured German early pattern body armour. Pilkem Ridge. July 1917. (IWM Q 2636)

Images of German soldiers as they saw themselves. Postcards produced for friends and relatives at home.

Column of 1st Guards Regiment with camouflaged helmets during the retreat. September 1918. (IWM Q 55403)

■ Young German prisoners wearing Model 1916 /17 steel helmets. They were captured during the German Spring Offensive. March 1918. (IWM Q 10845)

■ A German soldier with a square dip visored Model 1916 helmet. Western Front. (IWM Q 23760)

A dead German rifleman from the 203rd Regiment. Note that he wears a crudely cut hessian cloth cover tied down onto his helmet with thick twine. Near Bray. August 1918. (IWM E(Aus) 2955)

German troops wearing model 1918 cut-out steel helmets. Western Front 1918. (IWM Q 47997)

■ (Left) A Canadian soldier holds a badly damaged German Model 1916/17 helmet. (IWM Q 983) (Right) Lt. General Cosgrove shows General Currie a battered German steel helmet recovered from the front. (IWM CO 1624)

■ Canadians pile up discarded captured German steel helmets after the Battle of Amiens. August 1918. (IWM CO 3034)

BRITAIN

John Leopold Brodie's original 1915 Type B design for a helmet. Made of magnetic steel, this example has the Royal Artillery insignia painted onto the front and the 58th Division to the rear. Note battle damage, Hawkes & Co private purchase liner and early apple green paint to interior. Only 150 of these helmets are thought to have been produced.

Brodie's 1915 second prototype (magnetic) steel helmet (Type A).

Early issue six-tongued American cloth liner for Type A and B helmets with two-piece buckle chinstrap.

■ Brodie's 1915 War Office Pattern (magnetic) steel helmet adapted from the Type A prototype.

■ Early issue six-tongued American cloth liner also incorporated into the War Office Pattern helmet with two-piece buckle chinstrap.

■ Regulation 1915 issue War Office Pattern (non-magnetic) steel helmet.

■ Interior of the 1915 issue helmet fitted with the re-designed liner for the later 1916 Mark I helmet. Due to wear and tear, earlier model liners were regularly in need of replacement. Large stockpiles of earlier helmet parts meant that the old and new were often used in the same helmet.

■ Model 1916 (Mark I) with a formed metal rim added, produced in three sizes.

■ Re-designed liner for the regulation 1916 Mark I helmet, produced in several sizes with an adjustable drawstring net and a cushion ring of rubber. The chinstrap is the new sliding buckle type.

■ The early pattern chinstrap prong buckle detail (left) and the later pattern sliding buckle (right).

■ The early pattern chinstrap attachment to the shell (left) and the later pattern (right).

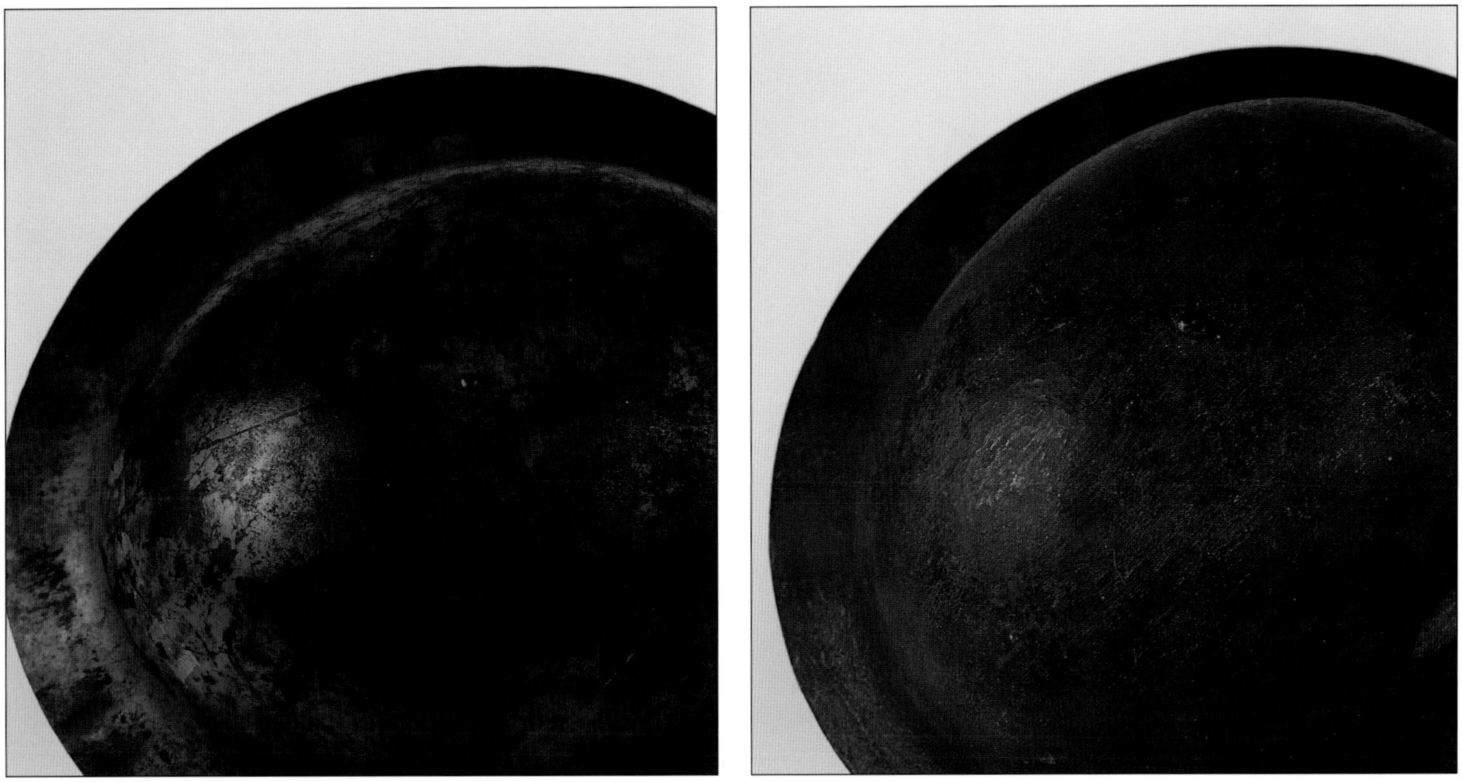

■ (Left) The standard copper rivet which attaches the liner to the shell. (Right) A private purchase screw down variation.

■ (Left) The early issue 1915 helmet rim detail showing manufacturer's stamping. (Right) Later Mark I helmet rim with maker's stamping and added mild steel rim.

■ Regulation 1915 War Office Pattern helmet as used by Canada's Fort Gary Horse (FGH) stenciled in white onto the early apple green colour.

■ Regulation 1915 War Office Pattern helmet belonging to Sapper A. Saul, Royal Engineers with 40th (New Army) Division's cockerel insignia painted onto the front.

■ Regulation 1915 War Office Pattern with the Machine Gun Corps insignia painted onto the front.

■ Regulation 1915 War Office Pattern with the Scottish 15th Division insignia painted onto the front.

Regulation 1915 War Office Pattern shell with private purchase liner having the 17th Hussars insignia painted onto both sides.

Regulation 1915 War Office Pattern helmet with the King's Own Yorkshire Light Infantry insignia painted onto the front.

Regulation 1915 War Office Pattern helmet with a heavy textured finish and Royal Fusiliers insignia painted onto the front. Shown below is the reverse with the 58th London Territorial Division insignia.

Two examples of the regulation 1915 War Office Pattern helmet carrying the painted insignia of the 34th (New Army) Division "chequer board".

Brodie's 1915 War Office Pattern steel helmet with the badge of the Leicestershire Regiment stenciled in red to the front. This helmet belonged to Major Tyler MC of the 7th Battalion.

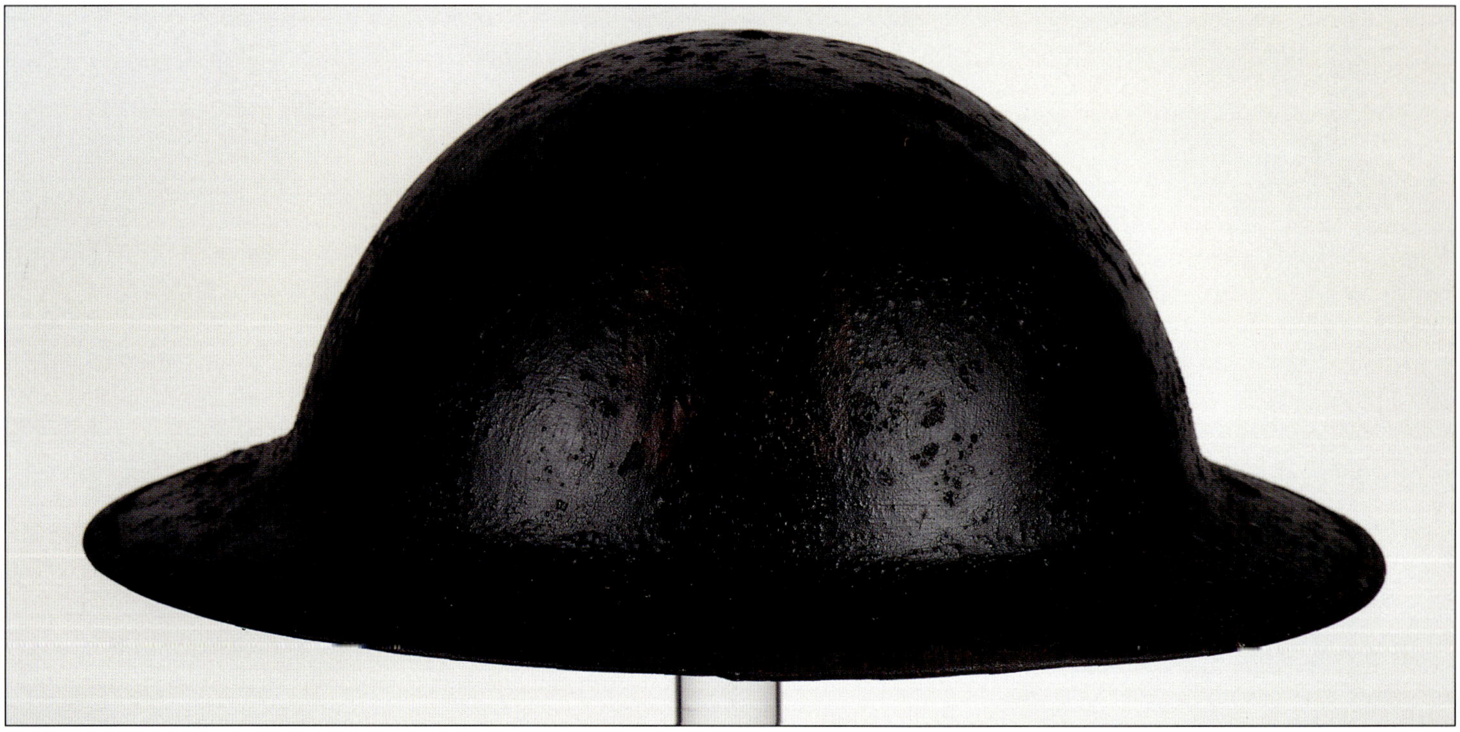

Brodie's 1915 War Office Pattern helmet shell with an unknown variant leathercloth liner and linen chinstrap.

Mark I (rimmed edge) with insignia of the King's Royal Rifles painted onto the front.

■ Mark I (rimmed helmet) with insignia of the 24th Division painted onto the front.

■ Mark I (rimmed helmet) with insignia of the 4th East Lancashire Regiment painted onto both sides.

Mark I (rimmed helmet) with the 2nd Dragoons Royal Scots Greys cap badge soldered onto the front.

Mark I (rimmed helmet) with the Manchesters regimental badge soldered onto the front.

■ Mark I (rimmed helmet) with insignia of the Royal Horse Guards cap badge soldered onto the front.

■ Mark I (rimmed helmet) with insignia of the Tank Regiment soldered onto the front.

■ Regulation 1915 issue War Office Pattern shell with private purchase liner and East Lancashire Regiment transfer on the front.

■ Privately produced magnetic helmet shell with Royal Garrison Artillery insignia painted on the front.

■ Regulation 1915 issue War Office Pattern shell with private purchase liner and 36th Division (New Army) Red Hand of Ulster insignia painted on the front.

■ Mark I (rimmed helmet) shell with private purchase liner and "clubs" card suit insignia painted on one side, thought to be 141st Brigade, 47th Division (the card suits identify particular Battalions).

Regulation 1915 issue War Office Pattern shell with private purchase liner made by Herbert Johnson, 38 New Bond Street, London and 3rd King's Own Hussars cap badge soldered onto the front and painted insignia to both sides.

Regulation 1915 issue War Office Pattern shell with Hawkes & Co private purchase liner and chinstrap.

Lightweight aluminium parade shell with a white leather Hawkes & Co private purchase liner.

Mark I (rimmed) helmet shell with rough sand texture finish and a Thomas White & Co, Union Street, Aldershot private purchase liner.

Regulation 1915 issue War Office Pattern steel helmet shell with a Walter Barnard & Son private purchase liner.

■ Hawkes & Co private purchase liner.

■ Private purchase liner variation by Lincoln Bennet & Co of Sackville Street, Piccadilly, London.

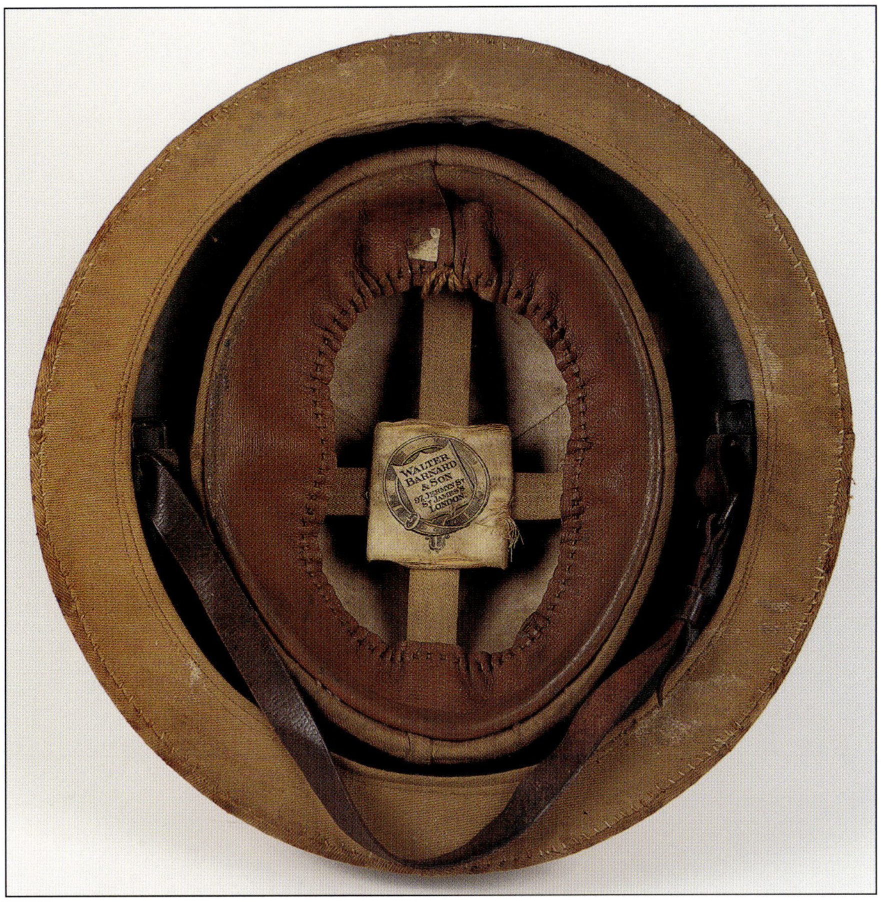

■ Walter Barnard & Son private purchase liner.

■ Woodrow private purchase liner variation (interior of Lord Hastings' helmet, see page 212).

Mark I (rimmed) helmet with standard issue liner and six-piece sacking cover with medal ribbon used as a unit flash sewn onto one side, identifying it as Northumberland Fusiliers.

Private purchase helmet with liner made by Woodrow of 46 Piccadilly, London and five-piece canvas cover. This helmet belonged to Lord Hastings (see pages 171 and 212).

■ Officer quality two-piece canvas helmet cover with North Staffordshire Regiment cap badge affixed.

■ Four-piece canvas helmet cover with tropical neck shade.

Private purchase helmet with liner made by Walter Barnard & Son. It has an officer quality four-piece canvas cloth cover. Note the metal sprung cushioning across the forehead section of the liner.

General officer's privately purchased helmet and four-piece cloth cover with metal cap badge sewn onto red cloth band and Hawkes & Co liner.

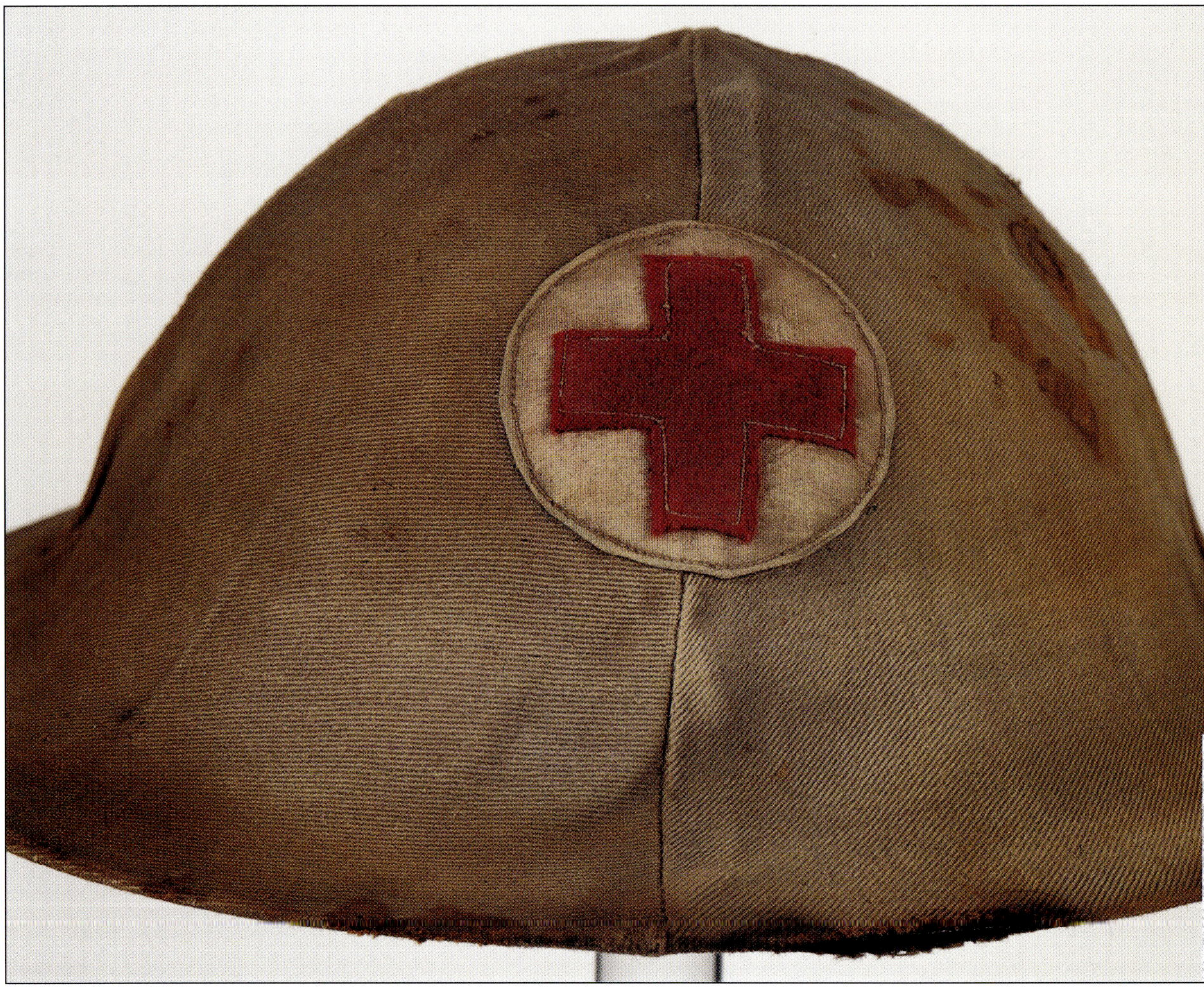

Helmet and cloth cover belonging to Victoria Cross winner Private R.G. Masters, 141 Field Ambulance ASC. The cloth cover is a four-piece construction with the Red Cross insignia sewn onto a white cloth patch. Private Masters won his award for repeatedly collecting the wounded under fire in his ambulance on 9th April 1918 near Bethune.

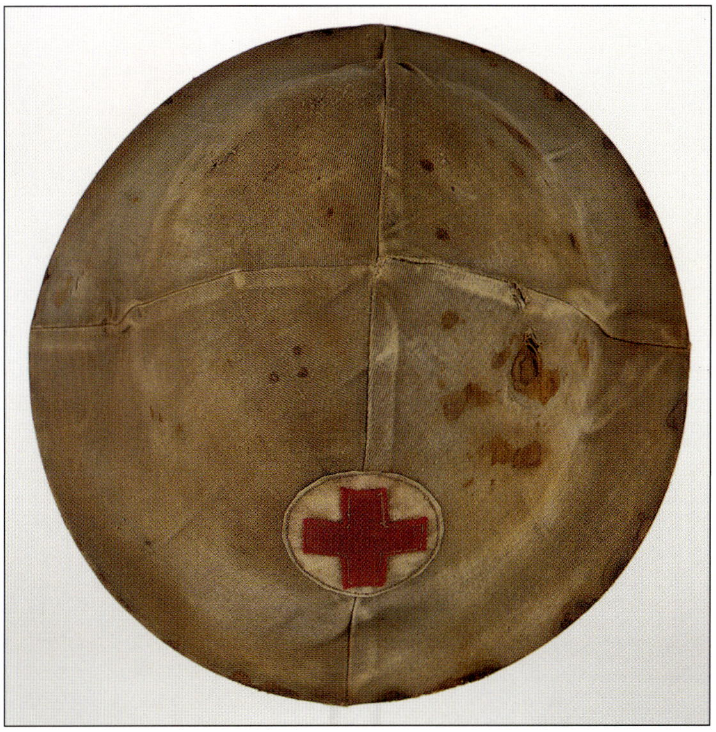

P.T.E. R.G. MASTERS
141. FIELD. AM.
A S C

Private purchase mild steel fluted helmet. Made by Henry Heath Limited 105-107 Oxford Street, London. Very similar to the helmets bought by the Portuguese for their troops on the Western Front. These types of mild steel helmets were later banned by the British authorities as they were considered worse than wearing no helmet at all. They would often fragment when hit and cause greater damage to the wearer. It carries the cap badge of the West Riding Regiment with the tongue of the badge inserted into a hole in the helmet.

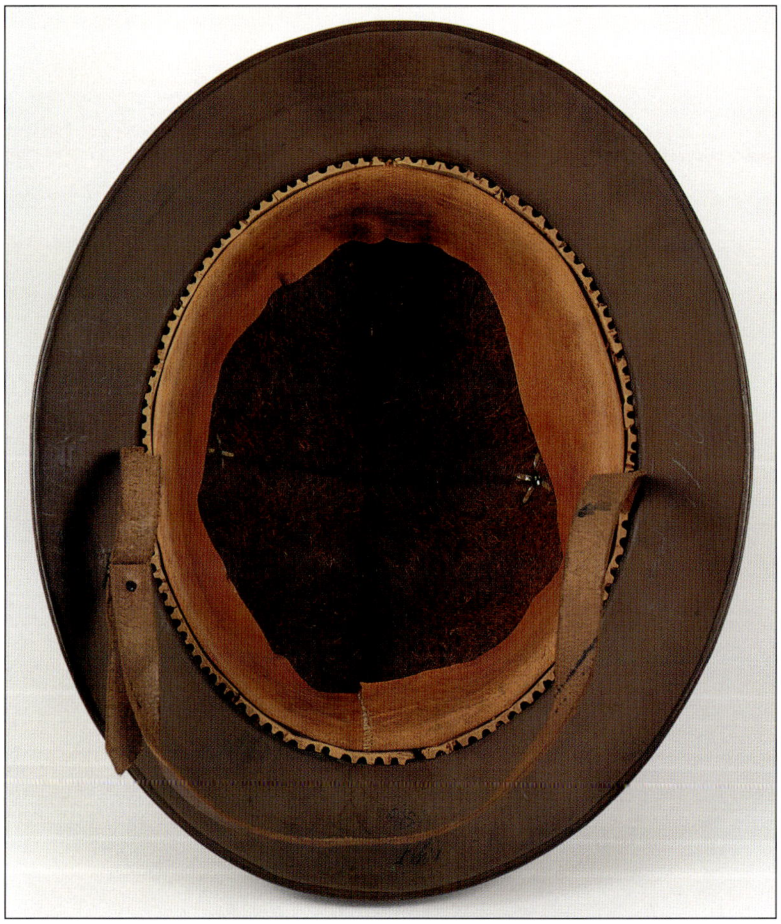

A variation of private purchase mild steel fluted helmet with leather band sewn through holes around the base of the bowl.

Another variation of private purchase mild steel helmet with leather band. This version is smooth rather than fluted and has a leather tongued liner.

A leather carrying case for the steel helmet, probably for the cavalry.

A variation of leather carrying case. This example owner-identified as R.F. Symonds.

Tankman's armoured face mask worn to protect crews against splinter and bullet splash.

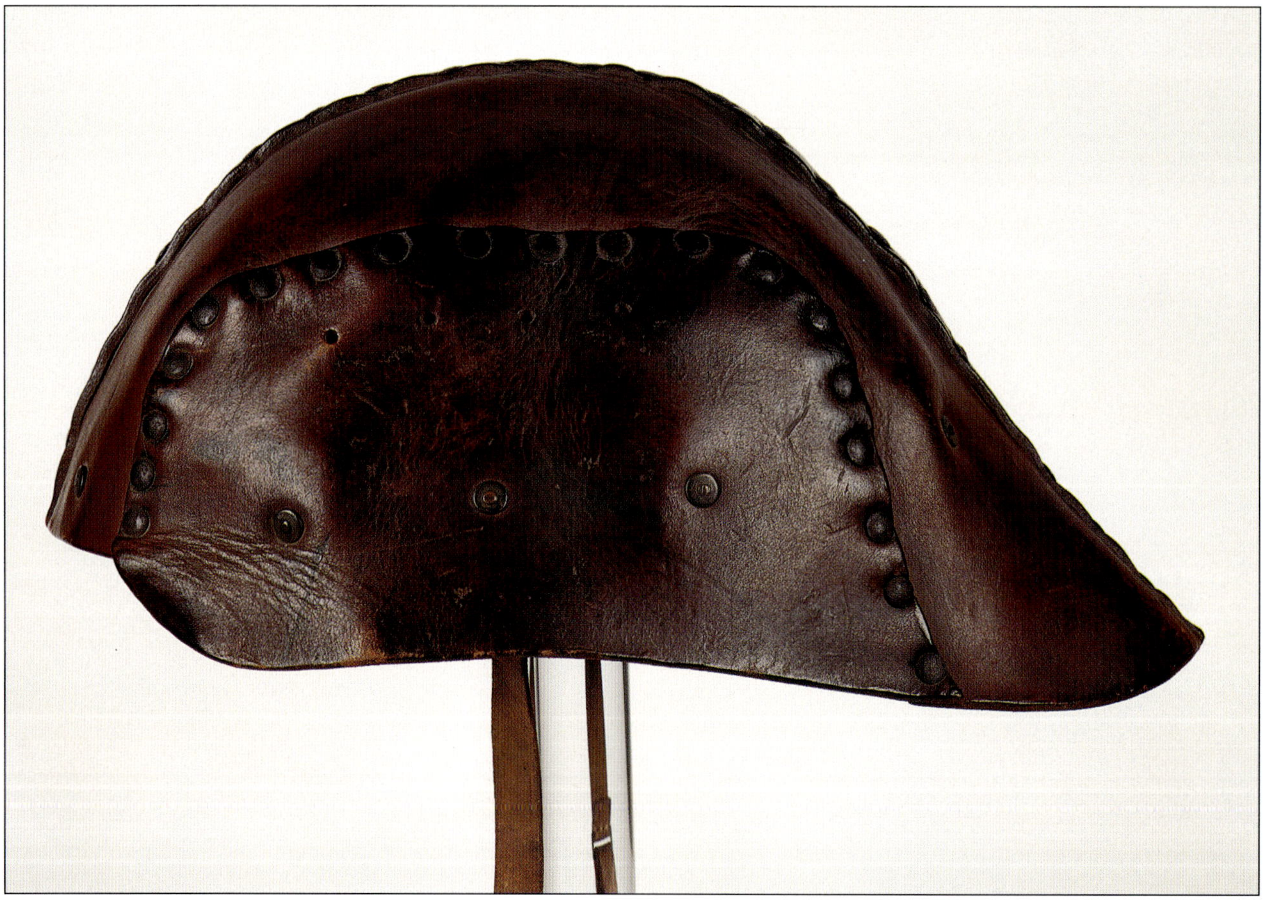

The leather crash helmet for tank crews. Issued in 1916 to prevent concussion inside a moving tank. They were soon withdrawn as they were thought to be too similar to the German helmet.

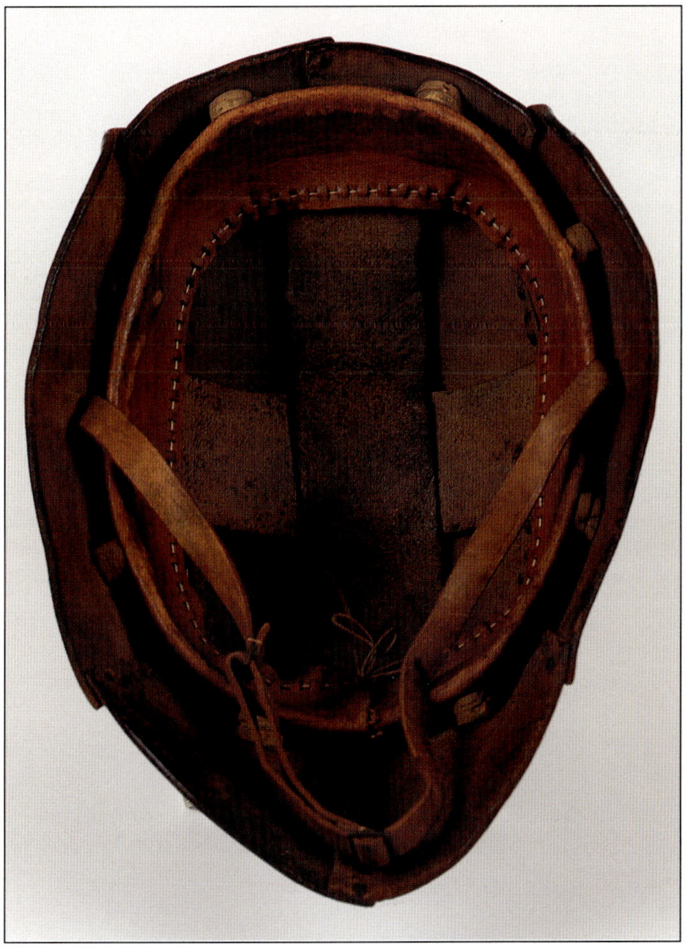

The Cruise Mark I helmet with face veil. This example is attached to a Brodie's 1915 War Office Pattern helmet with an early issue six-tongued American cloth liner. The chain mail curtain hangs from a rail and can be partially pulled across allowing the wearer to sight his rifle. It can also be pulled up over the helmet brim and hooked onto the horns either side of the bowl. Troops generally found this configuration an annoying distraction and often removed the curtain. These "unfrocked" helmets can still be found. Cruise helmets are named after their inventor Captain Richard Cruise, Royal Army Medical Corps and date from 1917.

The Cruise Mark II helmet and face veil. This later improved version has a central nasal bar which was attached to the chain veil and can be flipped up over the helmet rim. It is not entirely clear whether it was actually issued to troops during the Great War, but some were subsequently issued to police units during the Second World War, painted dark blue with "POLICE" painted in white across the front bowl.

All demobilised British troops were allowed to keep a helmet as a souvenir. This Mark I helmet (with rim removed) has been painted by its owner: Private S.H. Moore 1/6th Gloucesters, serial number 36998.

1/6TH Glosters
36998
33rd Division

1919
1914
1/6TH Glosters
36998

Another fine example of a souvenired helmet. This Mark I (rimmed) helmet belonged to a Royal Engineers "N" Special Gas Company man and has a complete record of his service painted onto both sides of the bowl from: 8th August 1915 "Landed Le Havre" to Jan 14 1919 "Demobilised at Duddingston".

FRANCE
1915 1919
ROYAL ENGINEERS
Coy. R.E.

June-July '18 Mailly-Maillet, Aveloy
-Wood, Auchonvillers,
Beaumont-Hamel
Aug. '18 River Ancre Offensive
Albert-Bapaume Offensive
Bapaume-Cambrai
Cambrai - Le Cateau
Sambre Canal, Berlaimont.
"Armistice"
Demobilised at
Coy. R.E.

An early private purchase French acier trempé (tempered steel) helmet owned by Second Lieutenant L.N. Kindersley of the 15th King's Hussars. He was killed in action at Bourlon Wood on the 25th November 1917. It was not unknown for British officers to buy French helmets prior to the official issue of British steel helmets.

■ British prisoners of war. The attachment of the liner to the helmet shell was never very satisfactory. Here we see many of the men wearing their detached liners as hats. (IWM Q 24051)

(Left) Brodie Type B helmets being worn. In a dressing station at Morlancourt during the Battle of the Somme. July 1916. (IWM CO 200) (Right) With a rough hessian cover near Robecq. July 1918. (IWM Q 11078)

An officer (centre) with a canvas cover over a Brodie Type B takes particulars from prisoners during the battle of Courcelette. September 1916. (IWM CO 815)

■ Brodie's 1915 War Office Pattern helmets. (Left) On a postcard. (Right) Photo taken in March 1916. (IWM Q 500)

■ Brodie's 1915 War Office Pattern helmets. (Left) After action at the St. Eloi craters, worn by a sergeant of the Northumberland Fusiliers. March 1916. (IWM Q 498) (Right) A view of the early issue six-tongued American cloth liner by the side of the Albert-Bapaume Road. July 1916. (IWM Q 781)

■ Private purchase mild steel helmets. (Left) Worn by an Artillery Forward Observation Officer, during the Battle of Arras. Near Guemappe. April 1917. (IWM Q 5279) (Right) By members of the Red Cross "Air Raid Brigade" circa 1917. (IWM Q 106930)

■ (Left) An officer with his trophy. He is wearing a private purchase helmet and cover. Courcelette October 1916. (IWM CO 857) (Right) A General wearing another private purchase variation. (IWM Q 3044)

■ Brigadier General Gurstall (right) wearing a private purchase helmet without a cover and Captain Papineau (left) with his variation with fitted cover (see previous page, bottom left). Circa 1916. (IWM CO 476)

■ (Left) An officer of the Fort Garry Horse (FGH). (IWM Q 9131) (Right) A soldier from the 58th Division chats to a French infantryman, Bernaguesse, south of Barisis. (IWM Q 8701)

■ (Left) A soldier of the 1/4th East Lancashire Regiment in a sap-head. Givenchy. January 1918. (IWM Q 7265) (Right) A Royal Field Artillery (RFA) man in the ruins of Ypres. September 1917. (IWM Q 2912)

■ A mounted officer of the 17th Lancers with metal cap badge attached to his helmet. Cologne, during the occupation. June 1919. (IWM Q 7722)

■ Brigadier General Loomis and his staff (2nd Infantry Brigade) wearing a mixture of private purchase and regulation helmets with cloth covers and metal cap badges. November 1917. (IWM CO 2232)

■ Major General Cuthbert wearing a plain unadorned helmet and Brigadier General Armytage with his private purchase helmet with fitted cloth cover and metal cap badge attached. Some of his staff wear plain helmets while others have chosen cloth covers or painted insignia for their private purchase helmets. Near Thiepval Wood. July 1916. (IWM Q 1065)

■ (Left) A British soldier with a coarse hessian cover. (IWM Q 23839) (Right) Lieutenant Colonel Bontemps, CO 126th Regiment, wearing a hessian helmet cover with embroidered cap badge attached. Granezza. July 1918. (IWM Q 26861)

■ (Left) An infantry battalion commander wearing a hessian helmet cover with a felt tactical Maltese Cross insignia sewn onto the rear of his helmet and jacket. Fonquevillers. April 1917. (IWM Q 1640) (Right) HRH The Prince of Wales wearing a private purchase helmet and cover with metal cap badge attached. Granezza. July 1918. (IWM Q 26858)

British troops using their Brodie 1915 War Office Pattern helmets as washing bowls, the liners having been removed. Ovillers. September 1916. (IWM Q 1095)

■ Men of the New Zealand Medical Corps (NZMC) with German prisoners. Some men wear hessian covers with a stitched drawstring band around the base of the helmet bowl to achieve a better fit. (IWM Q 3572)

■ Officers and men of the 107th Canadian Siege Battery. The officer at left wears what appears to be a woolen knitted stretch cover while others wear canvas and hessian with a metal cap badge affixed. July 1916. (IWM CO 448)

Lieutenant Colonel Lord Hastings in his capacity as CO of the Kent Home Guard during the Second World War (1940). Pictured here wearing his First World War private purchase helmet and cover (see pages 171 & 174). (IWM H 15772)

(Left) The General Officer Commanding (GOC) 2nd Australian Division and his aide in a trench named "Ungodly Avenue". Both officers have pleated canvas helmet covers and an attached metal cap badge. (IWM E(Aus) 1494) (Right) A British soldier cleans his rifle after coming out of the line. He wears a canvas helmet cover with his metal cap badge attached. Beaumont Hamel. November 1916. (IWM Q 4557)

■ (Top left) A sergeant collects discarded helmets with covers after the attack. Somme. July 1916. (IWM CO 210)

■ (Bottom left) Captioned as battlefield survey work being carried out. The officer wears a quilted canvas cover. (IWM Q 114589) (Right) A British prisoner of war wearing a much-used hessian helmet cover. Western Front. (IWM Q 24047)

A wounded Irish Guardsman and German prisoner. The British soldier wears a Cruise Mark I helmet with chain mail visor. Battle of Pilkem Ridge. July 1917. (IWM Q 7826)

Mark I Cruise helmets being worn: (Left) An Irish Guardsman after the Battle of Pilkem Ridge. July 1917. (IWM Q 2628) (Centre) Another Irish Guardsman at Pilkem Ridge wearing German body armour. (IWM Q 2636) (Right) An officer in the trenches on Hill 63 during the Battle of Messines. June 1917. (IWM E(Aus) 470)

■ Men of the Machine Gun Corps fire at a German aircraft. A Cruise Mark I is being worn by one member of the crew. Battle of Arras. 1917. (IWM Q 5173)

■ (Left) An example of an "unfrocked" Cruise Mark I being worn by Captain E. Jordan of the Worcestershire Regiment. He is pictured here in conversation with HM the King at Doullens Station. March 1918. (IWM Q 310) (Right) A British Lieutenant wearing a French Adrian helmet without its usual arm-of-service metal insignia. He is about to fire a Hales No.3 Mark I rifle grenade. Western Front. (IWM Q 110506)

Captioned as the first official photograph of a tank going into action, the corporal wears the leather tankman's helmet. Supposedly taken during the Battle of Flers-Courcelette. September 1916 (now thought to have been taken before the battle, possibly during a training exercise). (IWM Q 2488)

Stills from the Imperial War Museum's film "The Battle of the Ancre", they show the leather tankman's helmet being worn by crew members. 1916.

A British tank crew examine a German anti-tank rifle captured by Canadians. The kneeling officer and crew member holding the rifle are both wearing the armoured face mask issued to tank crews (see close-ups below and pages 186-187). August 1918. (IWM CO 2972)

ALLIES

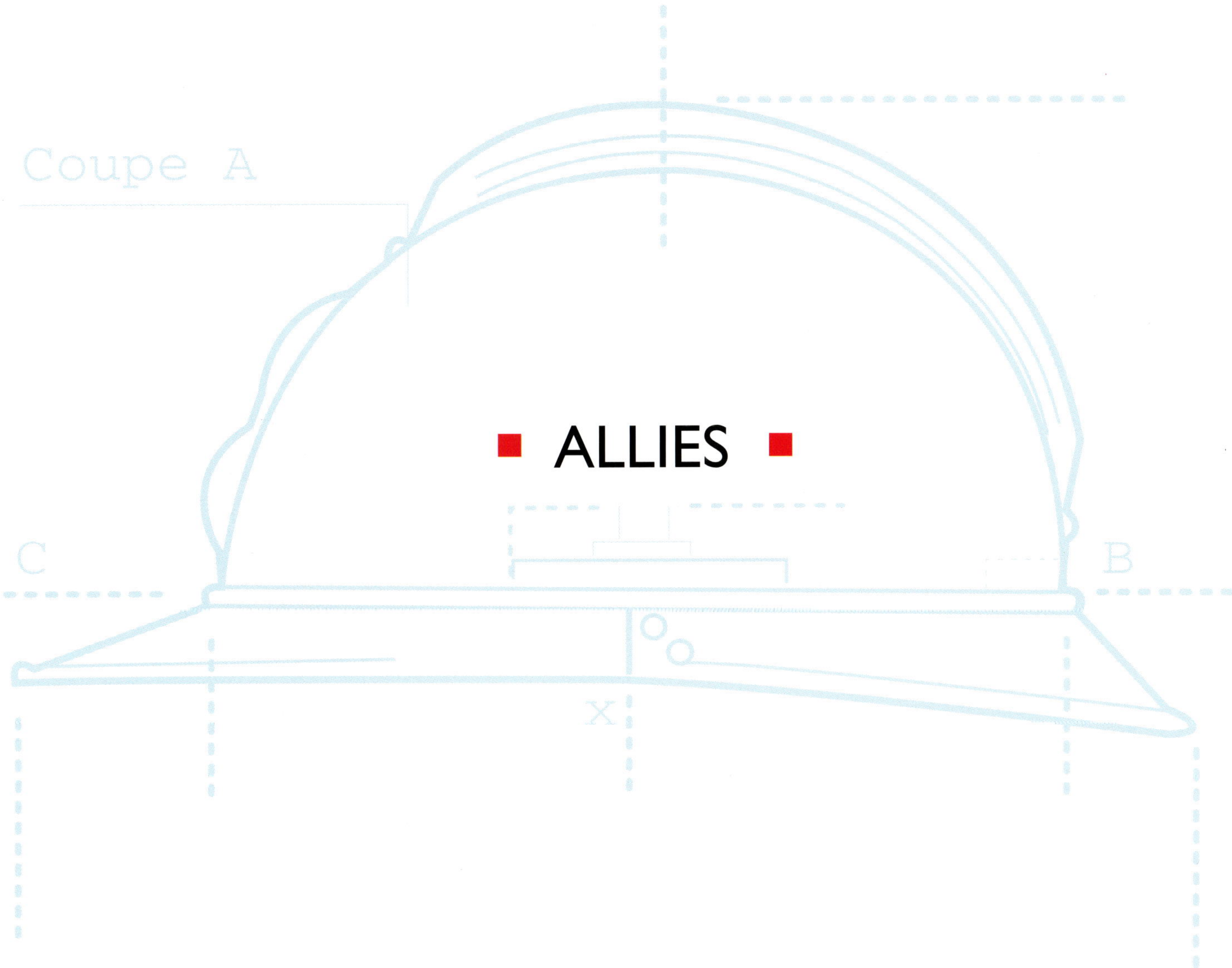

French Dragoons' helmet Model 1874 with its cloth cover. This type of helmet was largely out of service by 1915 and was superseded by a hybrid version incorporating features from the Adrian Model 1915 steel helmet.

■ French Model 1915 Adrian steel helmet. Designed by L'intendent (quartermaster) General A. Adrian, the shell is constructed from four separate pieces. This badge was used by the Infantry, Cavalry, Foreign Legion and Train Companies.

French Model 1915 Adrian steel helmet. This badge was used by the Medical Service.

French Model 1915 Adrian steel helmet. This badge was used by the Artillery and Tank Corps.

French Model 1915 Adrian steel helmet. This badge was used by the Engineers.

■ French Model 1915 Adrian steel helmet. This badge was used by the Tirailleurs, Zoaves and Spahis.

French Model 1915 Adrian steel helmet. This badge was used by the Colonial Infantry.

■ French Model 1915 Adrian steel helmet. This badge was used by the Chasseurs and African Riflemen.

Examples of various liner styles.

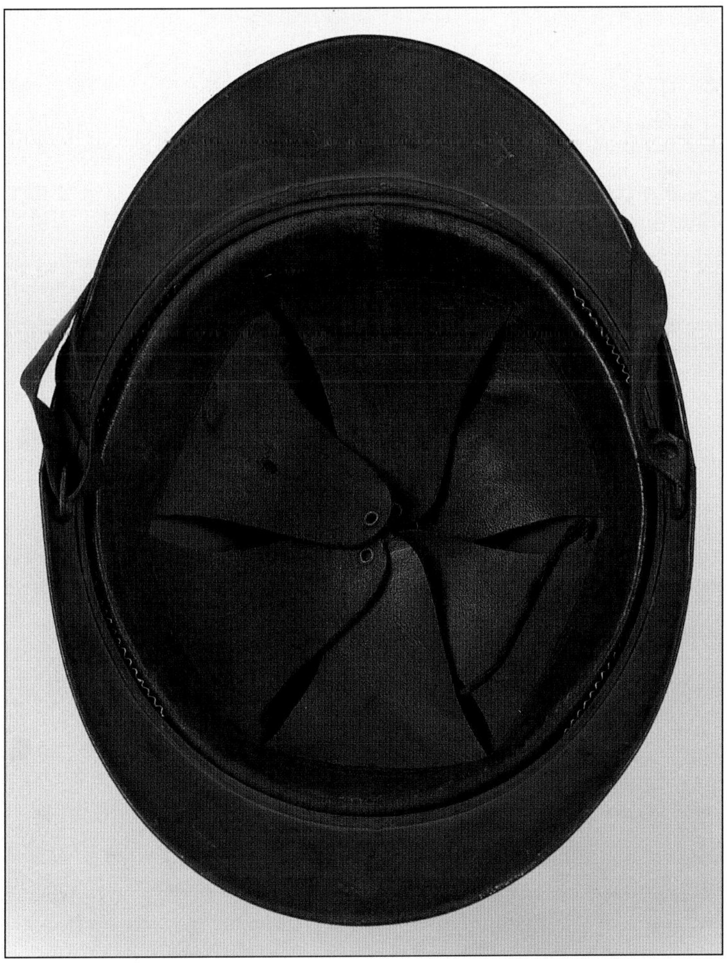

French Model 1915 Adrian steel helmet with Medical Service cloth cover.

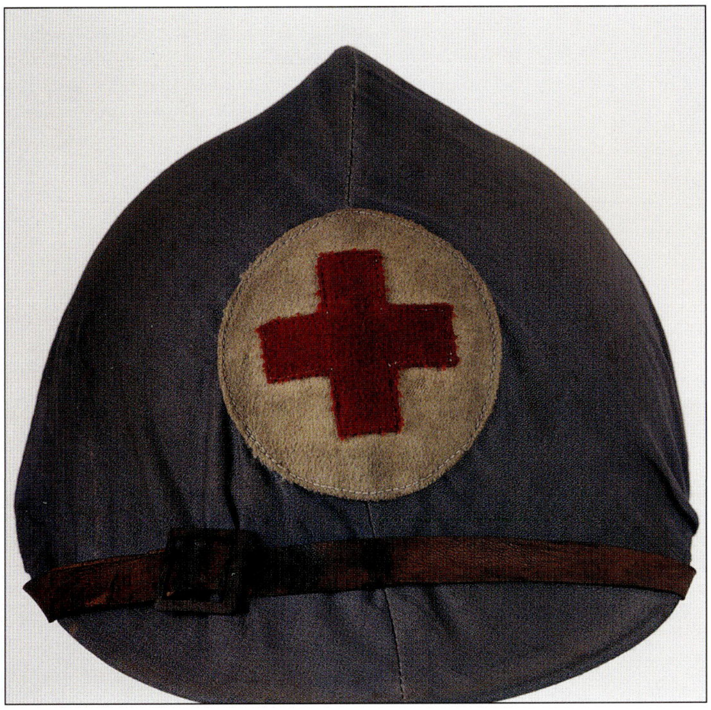

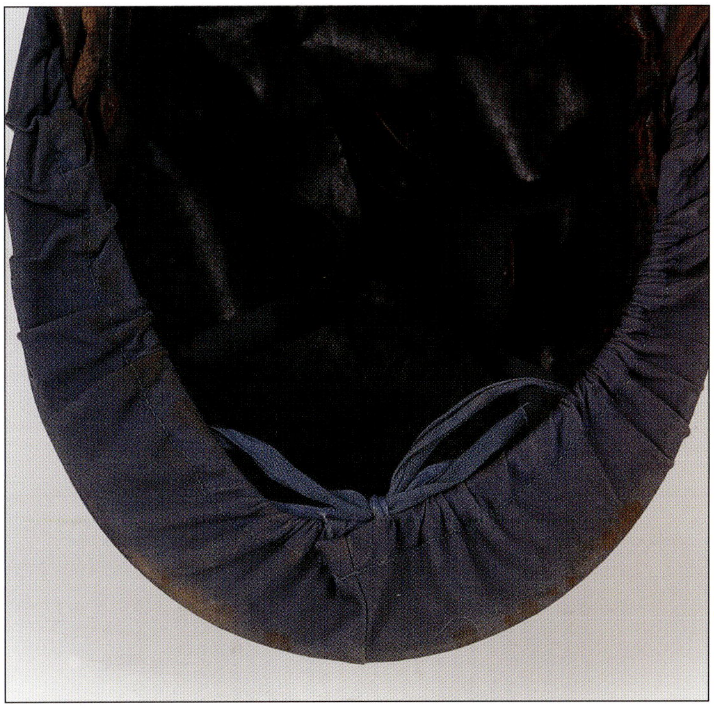

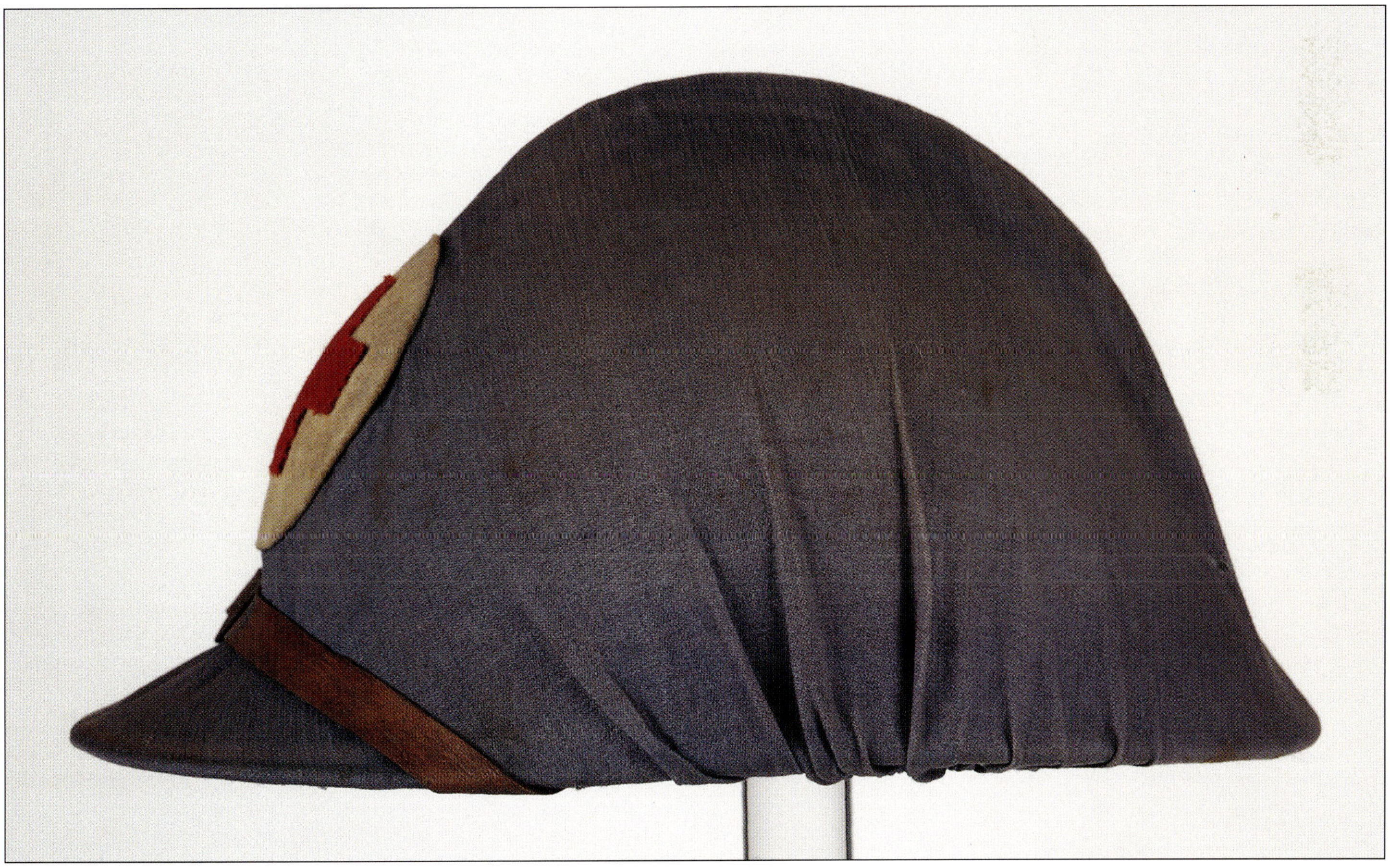

French Model 1915 Adrian steel helmet (known as the Artillery Special). The helmets were converted by hand in 1918 specifically for the use of tank crews. The helmet could be worn with the armoured face mask shown on page 186.

French Model 1915 Adrian steel helmet. Colonial Infantry badge. Showing battle damage.

French Model 1915 Adrian steel helmet. Infantry badge. Damage caused by a bullet or shell splinter.

■ French Model 1915 Adrian steel helmet. Taken as a souvenir and posted to "Mrs F. Robbins 1307 N Central Ave Phoenix Arizona" by "D. Robbins USS Carola Brest France".

French Model 1915 Adrian steel helmet with black painted chasseurs' badge. This helmet has had the veteran's brass plaque added and bears the inscription, "Soldat de la Grande Guerre 1914-1918". Some carry no details, but this example is named to Sergeant Chambon.

■ Belgian Model 1915 Adrian steel helmet. Manufactured in France.

Belgian Model 1915 Adrian steel helmet. Manufactured in France.

Belgian Model 1915 Adrian steel helmets showing different paint finishes.

■ Romanian Model 1915 Adrian steel helmet. Manufactured in France and supplied to Romania in 1917. Supplied in blue paint finish, this helmet has been repainted, probably after the war.

Czechoslovakian Model 1915 Adrian steel helmet. Manufactured in France and supplied to the Czech Legion fighting in France in 1918.

■ Russian Model 1915 Adrian steel helmet. Manufactured in France and supplied to the Russian government for their troops fighting in France and Macedonia.

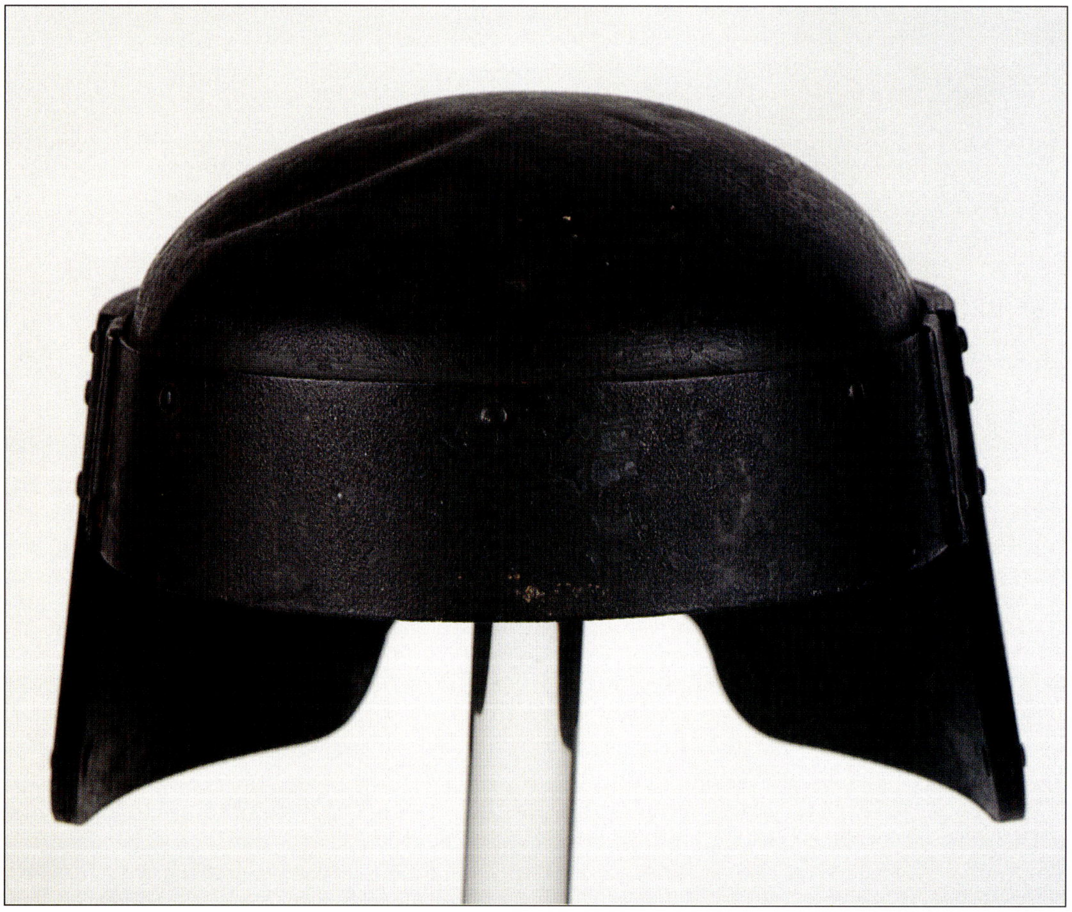

Italian armoured helmet (deep model) manufactured in Milan by the Farina company. The heavily armoured laminated plate can be worn to the front or rear. The liner was not attached to the helmet shell. This example is missing its chinstrap.

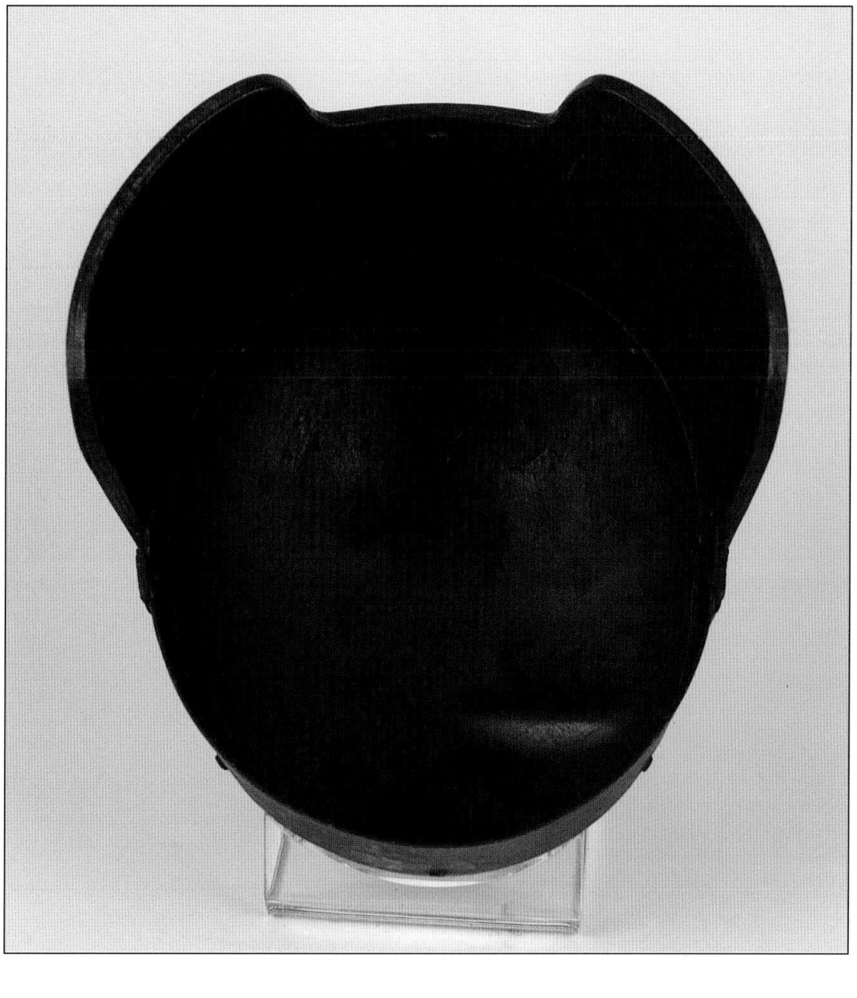

■ Italian armoured helmet (shallow model), also known as the trench model. Manufactured by the Farina company. This example is missing its chinstrap.

■ Italian body armour and helmet manufactured by the Farina company.

■ American Model 1917 steel helmet. The painted badge is that of the 37th Engineers (probably painted post-war as was often the case).

■ Interior view. Note the round rivets used to secure the chinstrap fittings indicating American manufacture.

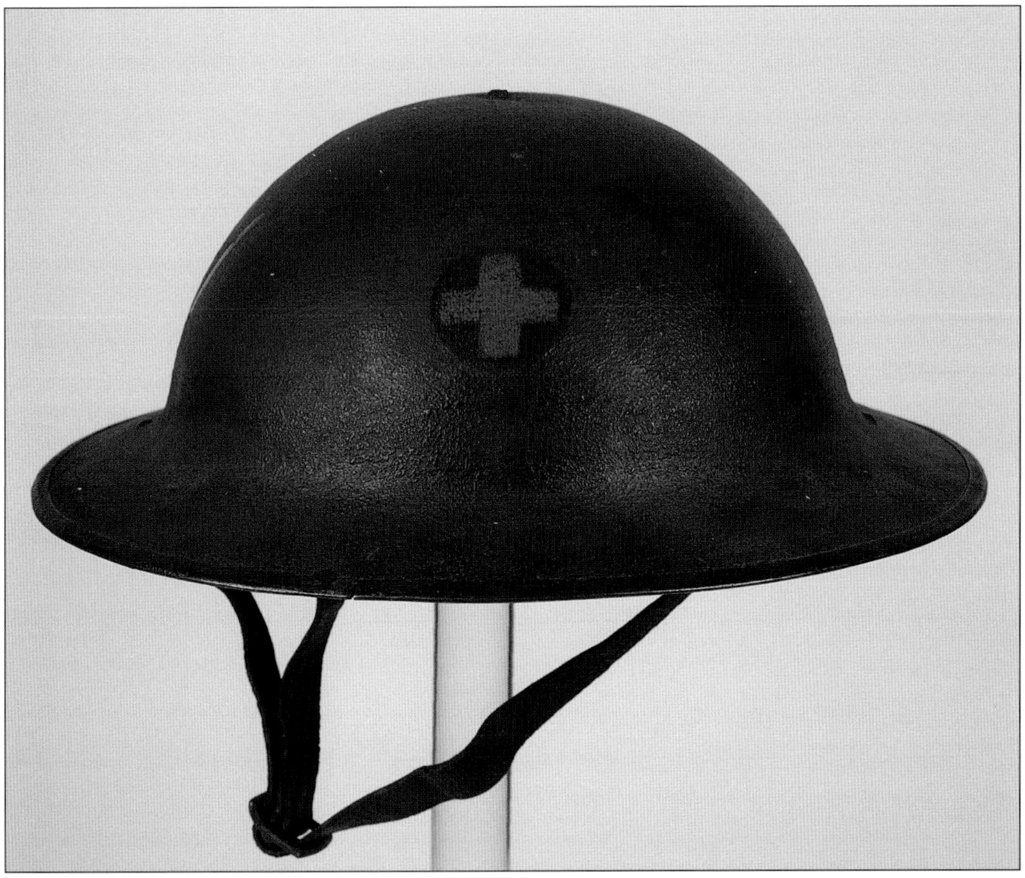

■ American Model 1917 steel helmet. The painted badge is that of the 33rd Division (probably painted post-war).

■ American Model 1917 steel helmet with camouflage finish and captain's rank bars. The paint style and worn condition suggest this helmet was painted during the war.

American Model 1917 steel helmet. Painted as a souvenir after the war. It bears the barred red arrow of the 32nd Division which arrived in France in February 1918. On the crown of the helmet is written "32nd DIV. 57th Brig. 120th F.A. les terribles". Around the brim is written "A.E.F. Haute Alsace-Aisne Marne-Fisme-Oise Aisne-Avocourt-Meuse Argonne". Attached to the chinstrap are the identity discs of Eugene Zielsdorf PVT. U.S.N.G. 2302414.

■ British Mark I steel helmet. One of the approximately 400,000 helmets offered by Britain to the American government. The painted badge is that of the U.S. 30th Division.

■ Interior view. Note the split pins securing the chinstrap fittings indicating a British-made helmet. This helmet is named to Pvt. Jas. G. Lewis-501379-Med. Dept.

Model 1917 Austrian steel helmet in regulation brown/khaki finish.

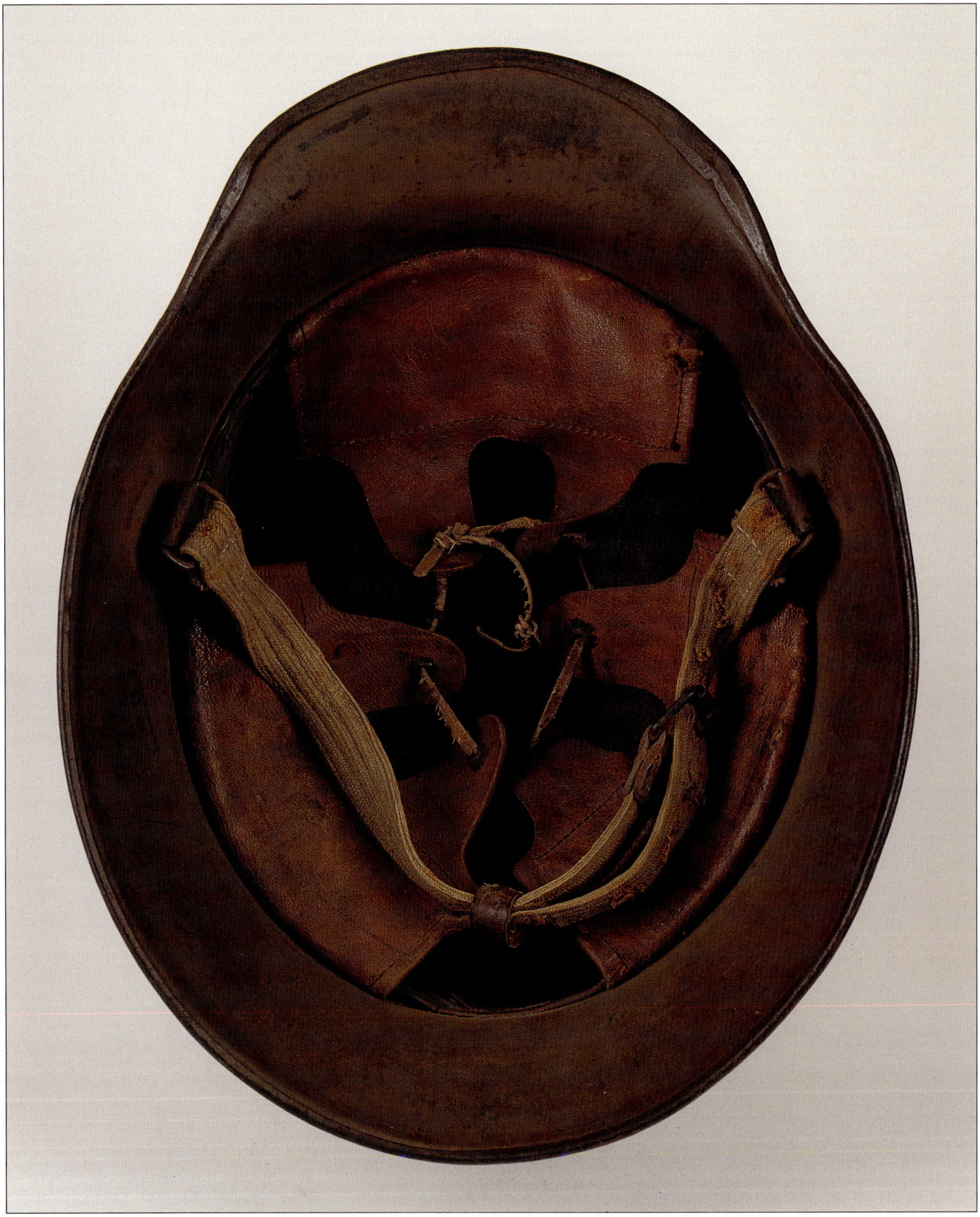

■ Model 1917 Austrian steel helmet interior. Note the similarity of the liner pads to the German version. The main difference being the addition of metal grommets for the drawstring.

■ Details of the strap attachment and maker's marks.

■ Details of the strap end and buckle. Note the typical leather trimming.

ALLIES

■ Model 1917 Austrian steel helmet in a typical German regulation camouflage finish.

Austrian armoured brow plate.

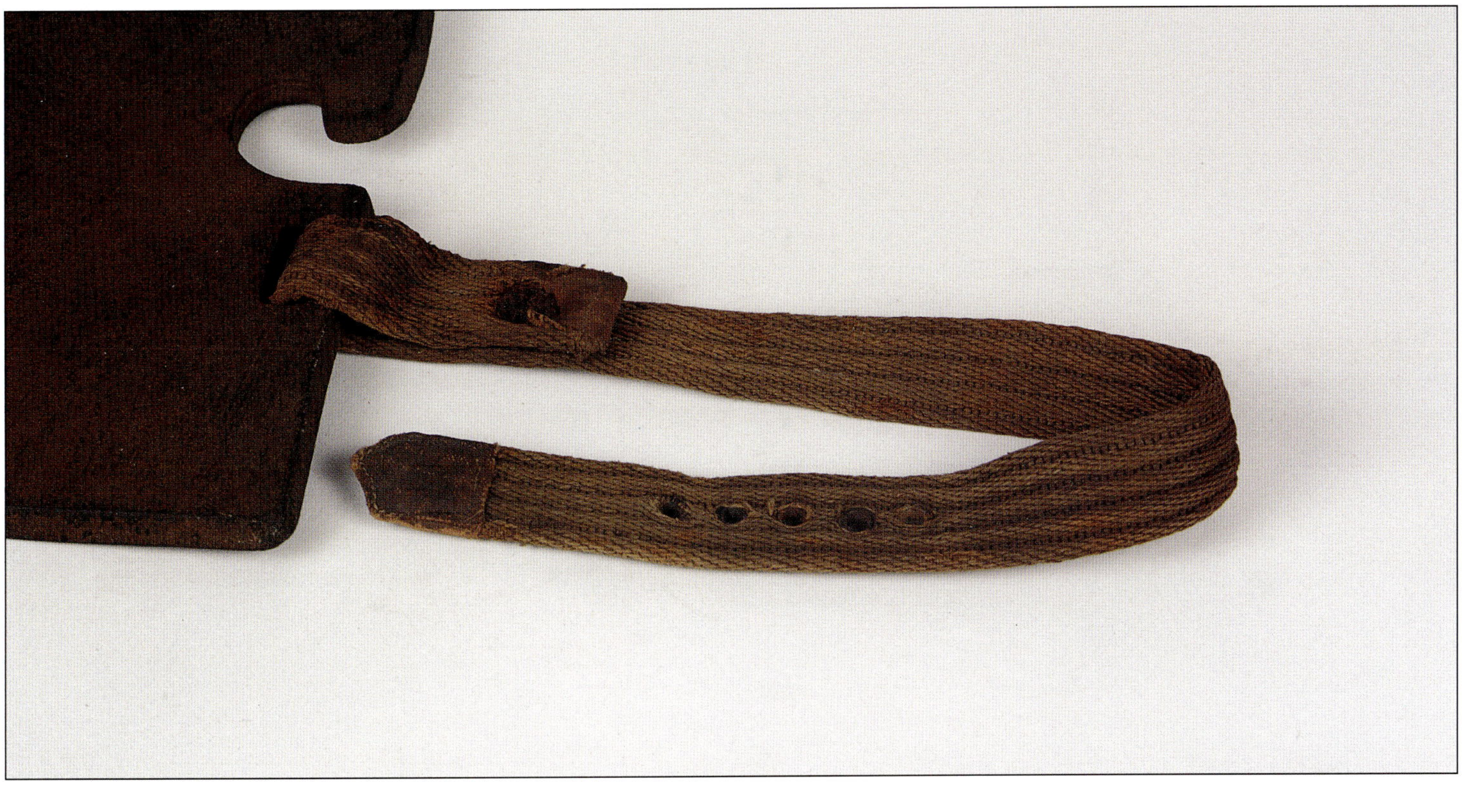

Details of the webbing straps, typical of the Austrian brow plate, the German version being leather.

Austrian Berndorfer helmet. Manufactured by Krupp Berndorfer Metalwarenfabrik in 1916/17. The single ventilation hole in the crown has a metal cover which also served as a fixing for the armoured brow plate.

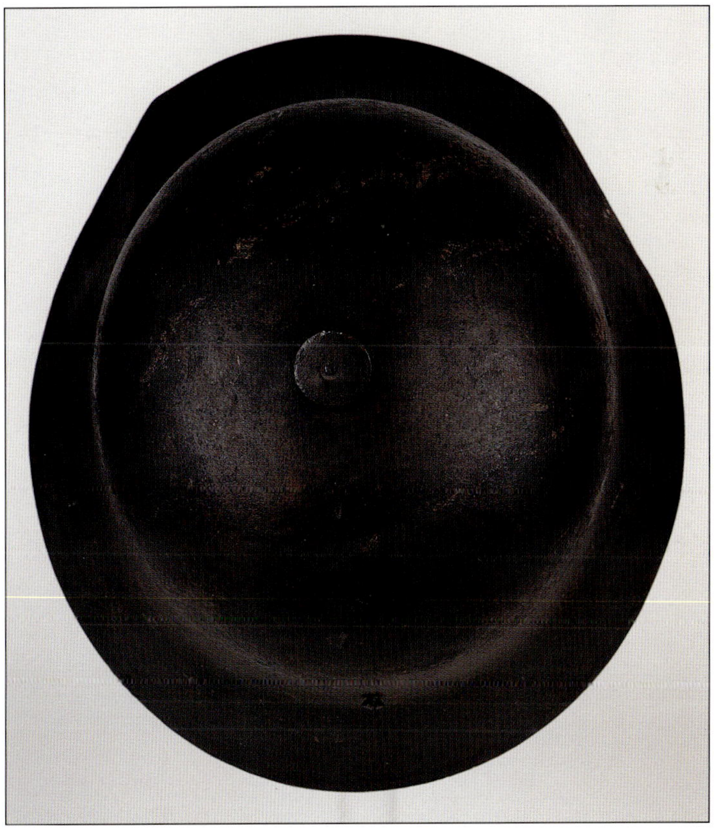

Interior view of the Austrian Berndorfer helmet. There are several styles of liner. This example has three leathercloth pads attached to a metal liner band. It is marked as belonging to the 38cm Howitzer Battery No.3.

■ Austrian Berndorfer helmet showing shell splinter damage. Note the carrying bracket at the rear of the helmet and the white painted rectangle to assist identification at night.

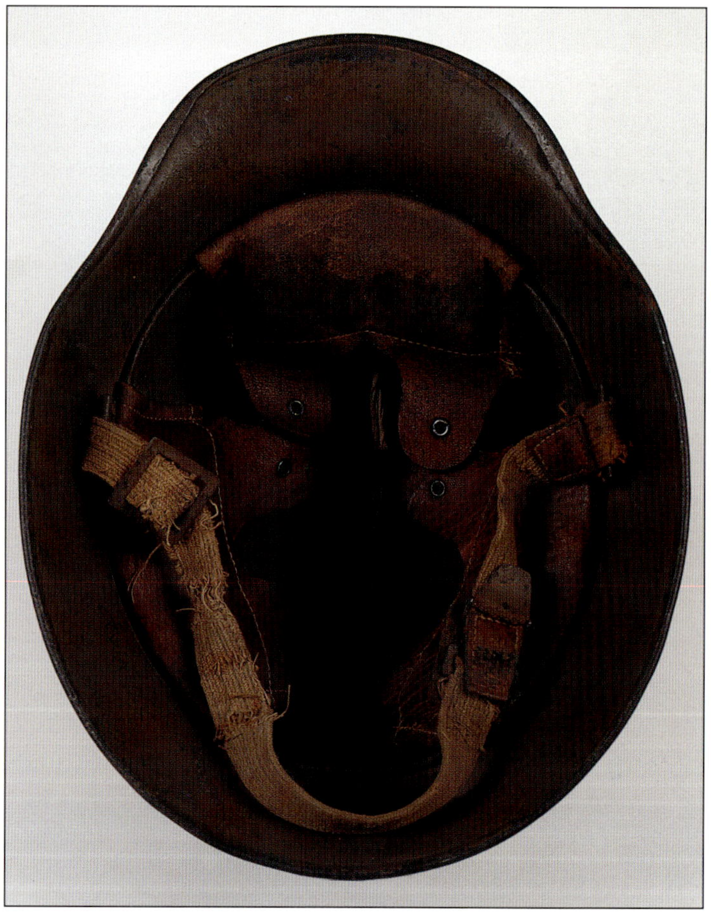

Hungarian Model 1917 steel helmet. Produced by Krupp Berndorfer Metalwarenfabrik. It differs from the Austrian Model 1917 in that the liner fixing rivets are smaller and positioned higher up on the bowl.

■ Model 1918 (visorless) helmet. 5,400 were produced by Germany for the Turkish government, but were never delivered. Used extensively by Freikorps units during the Sparticist uprising in Germany in 1919.

Model 1918 (visorless) helmet.

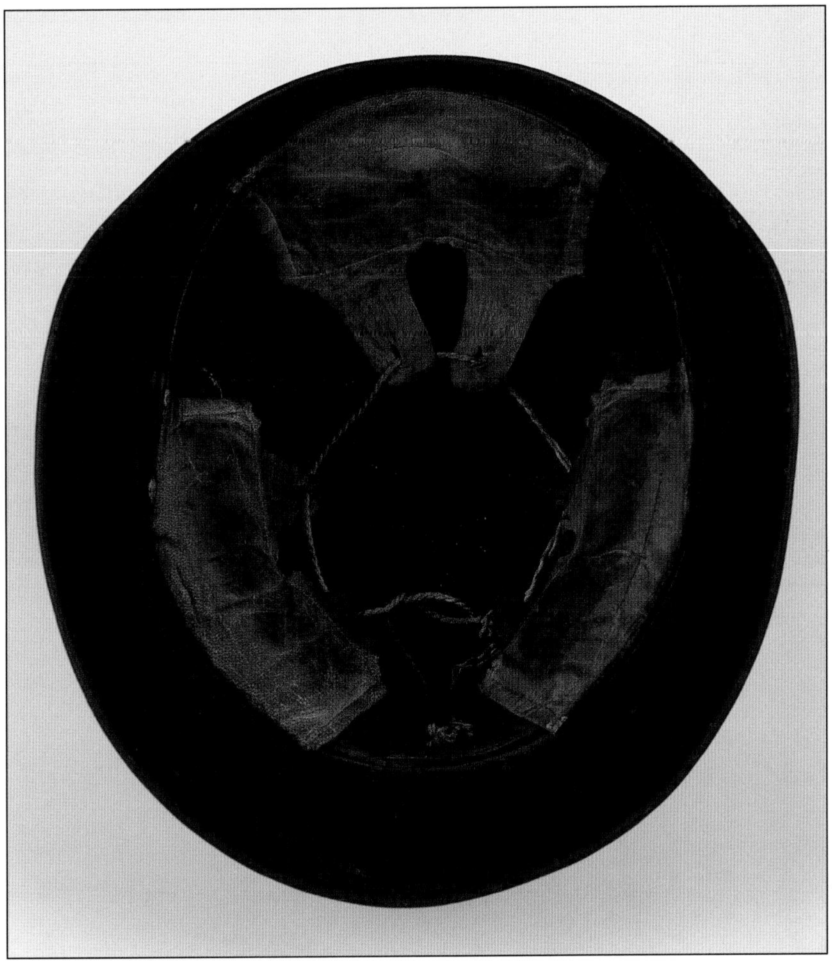

Model 1918 helmet (full visor variant) produced by Germany for the Turkish government.

Model 1918 helmet (full visor variant) produced by Germany for the Turkish government.

■ French infantrymen of the 117th Regiment wearing Adrian Model 1915 helmets in the early light blue finish.

French soldiers wearing Adrian Model 1915 helmets with a variety of badge plates, seen here after receiving British decorations. The soldier in the centre wears magazine pouches for the Chauchat automatic rifle. November 1916. (IWM Q 4530)

Mounted French troops (Chasseurs) with the appropriate insignia on their Adrian Model 1915 helmets. Near Domart. April 1918. (IWM Q 10866)

■ (Left) A French artilleryman shares wine from his canteen with British comrades. Pilkem Ridge. August 1917. (IWM Q 2741) (Right) F. Gargill Begg, CO of Section Sanitaire Anglaise 4. He wears an Adrian Model 1915 French helmet with an artillery badge which seems to have been highlighted in gold or silver paint. Vosges. 1915/16. (IWM Q 115318)

■ General Haking of IX Corps, seen here talking to members of the Lille Fireman's Band. The Germans had taken their copper-rich helmets and replaced them with captured British ones. The band appear to have added the Firemen's insignia to the front. October 1918. (IWM Q 9573)

(Left) General the Earl of Cavan. (Right) General Giazianin who is wearing an Adrian Model 1915 infantry helmet with metal Generals' stars attached. The chinstrap is the officers' braided type. Granezza. July 1918. (IWM Q 26856)

(Left) A French artillery officer wearing an Adrian Model 1915 helmet with officer-quality braided leather chinstrap. Bernagousse. March 1918. (IWM Q 8566) (Right) The French infantry officer on the left wears the braided chinstrap while the Chasseurs officer on the right wears the plain version. Granezza. July 1918. (IWM Q 26862)

■ French tankmen wearing Adrian Artillery Special helmets. They are riding atop their Renault tank which is in turn being transported by lorry. Aisne. 1918. (IWM Q 56445)

Belgian troops wearing Adrian Model 1915 helmets. September 1917. (IWM Q 2953)

A Belgian soldier working on his "Trench Art" in the September sun of 1917. (IWM Q 2960)

■ Russian troops await the order to enter Monastir after its capture from the Bulgarian and German forces. They wear the Adrian Model 1915 helmet with the badge of the Romanoff double-headed eagle attached to the front. November 1916. (IWM Q 32855)

Italian engineer troops wearing the (deep) Farina armoured helmet and body armour.

A wounded British soldier at a dressing station wears an Italian Farina helmet (reversed) over his service cap. Salonikan Front. (IWM HU 52413)

■ United States officers of the 305th Machine Gun Battalion consult with a British staff officer.They wear the Brodie style Model 1917 helmet. Watten. May 1918. (IWM Q 8844)

■ United States Marines wearing Brodie style Model 1917 helmets in the trenches at Breuvannes. February 1918. (IWM Q 70181)

■ (Left) An American lieutenant with a metal badge affixed to the front of his helmet. Western Front. 1918. (IWM HU 76123A) (Right) An American machine gun crew fire a French-made Hotchkiss. Western Front. 1918. (IWM Q 57006)

■ An Italian soldier about to throw a grenade, wearing a French-made Adrian Model 1915 helmet. (IWM Q 65142)

■ Hungarian mortar troops wearing German Model 1916/17 helmets. August 1918.

■ Hungarian troops wearing a mixture of Austrian Model 1917 and German Model 1916/17 helmets. June 1917. Note the slung grenade bags.

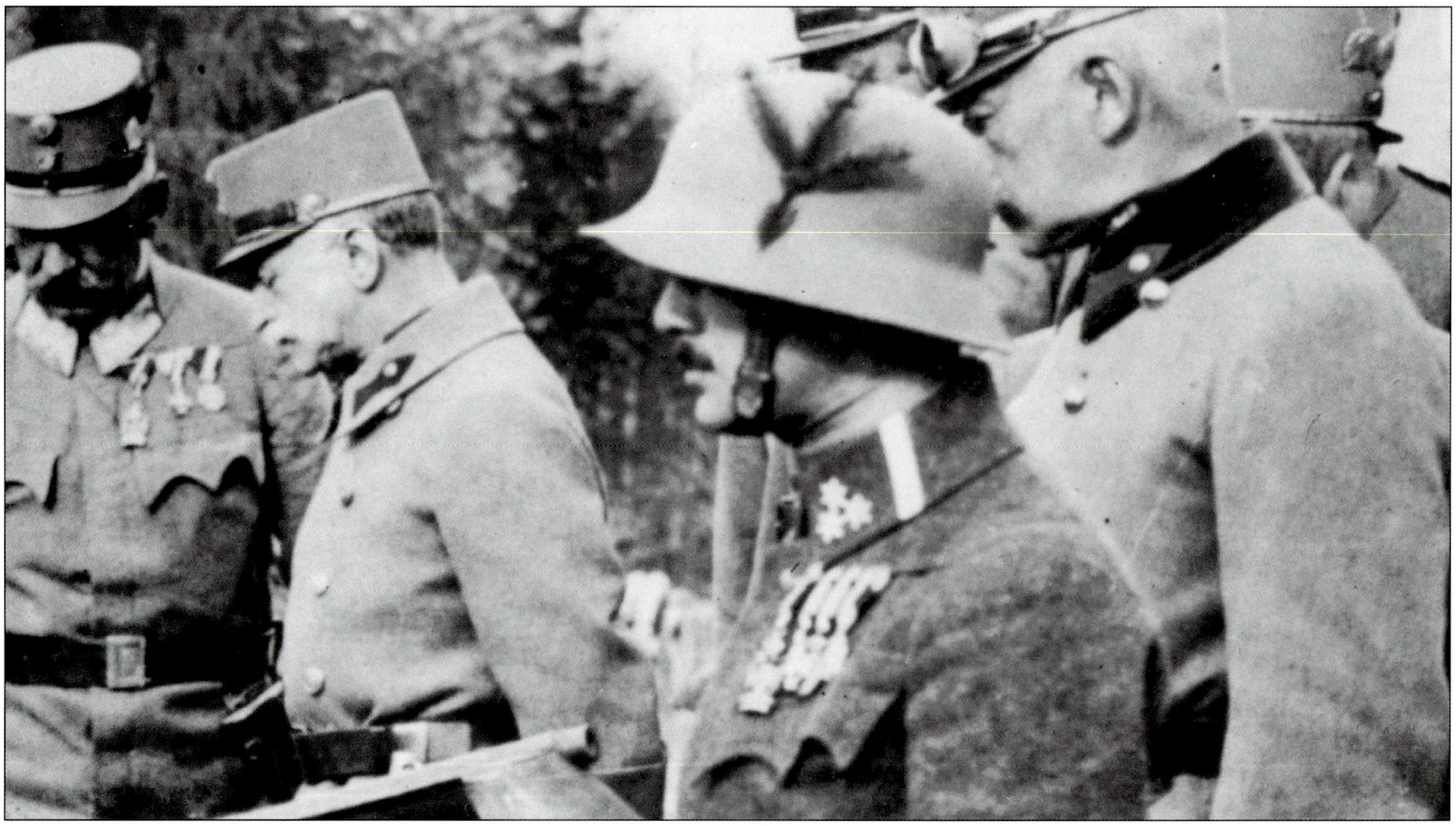

■ An Austrian prisoner of war wears a Berndorfer helmet with leather chinstrap. July 1918. (IWM Q 26805)

■ An Austrian officer wearing a Berndorfer helmet. Austrian helmets were sometimes decorated with three oak leaves in summer or in winter, three sprigs of fir. (IWM Q 64322)

■ A Turkish officer on his horse wears the German-made Model 1918 full visor helmet. It was thought that the removal of the visor was to allow Germany's Muslim allies to pray more easily. It is quite possible that the full visor version mimicked the soft Turkish service cap pictured here in inset, worn by the Kaiser during a visit to the Turkish Front. (IWM Q 23731)

■ Turkish storm troops (note grenade bags) wearing Model 1918 full visor helmets, being inspected by a German officer wearing a Turkish style fur cap. (IWM Q 80044)

Suggested Reading

The History of the German Steel Helmet 1916-1945 by Ludwig Baer

Steel Pots by Chris Armold

Les Casques De Combat volumes 1, 2 & 3 by Charles-Lavauzelle

Combat Helmets of the World by Paolo Marzetti

Les Casques De L'Armée Francaise 1915-1945 by Frederic Bruyelle, Renaud Mansuy and François Raquillet

Storia Dell'Elmetto Italiano by General Franco Angioni

Militaria Magazine (Histoire et Collections) A number of excellent articles. Attention is particularly drawn to issues 40, 54, 76, 88, 91, 98 & 101 which contain articles on British helmets of the First World War by Marcus Cotton

Helmets and Body Armour in Modern Warfare by Bashford Dean

Stahlhelm by Floyd R. Tubbs

The Pickelhaube volumes 1 & 2 by J.A. Bowman

Armi Ed Equipaggiamenti Dell'Esercito Italiano Nella Grande Guerra 1915-1918 by Nevio Mantoan

French Military Helmets of the XXth Century Roland Hennequin

The above list is in no way intended to be exclusive and is certainly not intended as a comment on the value of any book we have omitted to mention.

The Authors
August 1999

NOTES

NOTES

NOTES

NOTES